BEYOND BETA

BEYOND BETA

Other Continuous Families of
Distributions with Bounded Support
and Applications

Samuel Kotz • Johan René van Dorp

The George Washington University, USA

World Scientific

NEW JERSEY • LONDON • SINGAPORE • BEIJING • SHANGHAI • HONG KONG • TAIPEI • CHENNAI

Published by

World Scientific Publishing Co. Pte. Ltd.

5 Toh Tuck Link, Singapore 596224

USA office: 27 Warren Street, Suite 401-402, Hackensack, NJ 07601

UK office: 57 Shelton Street, Covent Garden, London WC2H 9HE

British Library Cataloguing-in-Publication Data
A catalogue record for this book is available from the British Library.

BEYOND BETA
Other Continuous Families of Distributions with Bounded Support and Applications

ISBN 981-256-115-3

Printed in Singapore by World Scientific Printers (S) Pte Ltd

To my wife Greer, my daughter Madeline and
my parents Govert and Elizabeth
Johan René van Dorp

To Rosalie, Tamar, David and Pnina
Samuel Kotz

Preface

This monograph is devoted to the topic of univariate continuous distributions *on a bounded domain other than the beta distribution*. As early as 1919, E. Pairman and K. Pearson recognized the topic of continuous distribution on a limited range to be an important aspect of statistical theory and investigated estimation of their moments. However, even in the late nineties of the 20-th century only a relative few number of probabilistic models of this kind were available. Amongst them, the uniform, triangular and beta distributions are the most widely explored and used, interspersed by some "curious" distributions posed occasionally as problems or exercises. Other bounded continuous distributions were based on mathematical transformations of the normal distribution (with an unbounded domain) - the most wide-spread amongst them is still the Johnson S_B family of transformations introduced in 1949 (briefly discussed in Appendix B).

Aside from the latter system of distributions, a comparison with the multitude of existing <u>unbounded</u> continuous distributions developed in the 20-th century is striking and even somewhat puzzling. There are of course historical reasons for this discrepancy — the main one being that the basic origin of continuous distributions stems from the famous Pearson family (containing mostly unbounded distributions) which in turn is related to the most prominent continuous distribution — the so called Gaussian or normal (symmetric) distribution (likewise unbounded). This distribution has a multitude of names. Interested readers should consult <u>The History of Statistics</u> by S.M. Stigler (1986) or/and Hald (1990). Perhaps it is no coincidence that the latter distribution was used as basis for the Johnson S_B system of bounded distributions. It should be noted that the unbounded continuous distributions usually computationally have the property that estimation of their parameters does not pose serious problems since originally estimation procedures were developed with an eye on the unbounded families.

Even when dealing with a problem of a bounded nature, such as, e.g., the propensity of particular physical characteristics of the human body, unbounded distributions such as the Gaussian distribution have been used to describe the uncertainty about these characteristics. Situations are not limited to the "physical" problems. For example, in the world of finance logarithmic transforms of interest rates (typically a quantity between 0% and 100%) are combined with their one-step differences to arrive at an unbounded domain, allowing once again the use of the Gaussian distribution for describing uncertainty in these one-step differences. In cases similar to the one described above, authors usually emphasize that they are using the unbounded continuous distribution as an *approximation* to the actual bounded state of affairs. In our opinion such an approach is not quite necessary in the 21-st century when computational difficulties associated with estimation of parameters of a distribution do not pose problems any more and are easily overcome. At the very least, a more natural continuous distribution with a bounded domain should be available that describes the uncertainty of such bounded phenomena.

The only continuous bounded distribution discussed extensively in numerous textbooks on probability and statistics is the well known beta distribution or the Pearson type IV distribution, which is highly flexible and has been used — occasionally indiscriminately — for fitting of data stemming from various fields. This approach is usually empirical since the parameters of the beta distribution (in its original form) do not have a proper physical meaning and having fitted beta distributions, the fitted beta parameters themselves do not always provide a picture of the phenomenon generated by the data under scrutiny.

Moreover, like the normal distribution, the beta distribution is smooth. Whereas a "peaked" alternative for the normal distribution has been available for quite some time such as the Laplace distribution, a flexible "peaked" alternative for the beta distribution has been lacking until very recently. Smoothness of density curves may be an attractive mathematical property, but it does not always have to be dictated by the uncertainty of the phenomenon one is attempting to describe. In particular, financial data has been shown to exhibit "peaked" histograms and even displayed jump discontinuities. This fact prompted us to search for "alternative curves" on a bounded domain which:

(1) possess properties of the uncertain phenomenon one is attempting to describe,

(2) mimics the flexibility of the beta distribution and

(3) possess meaningful interpretation of parameters which permit to classify the data and draw practical conclusions.

Our initial efforts originated as early as 2000 and resulted in the construction of the *two-sided power (TSP) distributions* which cover almost all varieties of the beta distribution, enhances flexibility in the unimodal domain, possess the cumulative distribution function and quantiles in a closed form, and whose parameters provide information on the structure and properties of the distribution at hand. The distribution was independently introduced in a technical report in 1985 by B.W. Schmeiser and R. Lal.

The material available by now (2004) on bounded continuous distributions allowed us to compile a monograph consisting of 8 chapters containing predominantly novel contributions together with some neglected models. Indeed, while searching the literature, we have encountered a number of relatively obscure continuous bounded distributions, having attractive statistical properties which extend the realm of models that can be represented by bounded continuous data. A common thread amongst most of these distributions is that they can be viewed to be extensions of the triangular distribution. It would seem that these distributions were not sufficiently covered in the relevant periodical and monographic literature prior to the 21-st century. As an example, aside from the authors recent publications, no maximum likelihood procedure seems to have been available for the widely encountered three-parameter triangular distribution. It appears that only some ad-hoc estimation methods have been provided in distribution fitting software that do not contain detailed descriptions of their fitting procedures.

The initial Chapters 1 to 2 cover the triangular distribution and some of its pre 21-st century extensions, which we consider to be of substantial historical and of practical value. The text of Chapters 3 to 8 records — to a large extent — our own contributions to the post-20-th century alternatives to the beta distribution which are related to the TSP family possessing attractive properties not always shared by the beta distribution. The distributions presented in Chapters 3 to 7 are motivated by utilizing real world engineering and economic data. Some of the illustrative examples are elaborate and in Chapters 3 and 6 involve modeling of financial time series and in Chapter 7 deal with modeling of income distributions for African American, Hispanic and Caucasian (non-Hispanic) subpopulations in the U.S. revealing stochastic ordering amongst them. Examples in the remaining chapters are based on civil engineering and physics data.

We are not aiming to dethrone or demolish the time-honored beta distribution which will no doubt continue to have a secure place and plenty of meaningful applications in statistical, engineering, medical, econometrics and other applied literature, as indicated by the recently published handbook on the beta distribution by Gupta and Nadarajah (2004). Our intention is to enlarge the arsenal of univariate continuous bounded distributions and thus improve the description of those statistical models to which the newly proposed distributions may well be applicable. While in this edition we predominantly focus on the triangular distributions and its extensions, we hope to provide a more comprehensive treatment of univariate continuous distributions on a bounded domain in a future edition and emphasize the computer-oriented aspects.

We are grateful to a number of people for helping us to bring this year long project to fruition. In preparing this monograph we have benefited from the advise of Prof. T.A. Mazzuchi (The George Washington University), Prof. S. Nadarajah and as well as suggestions from the editors and referees of our own published papers on this topic. We are indebted to Dr. Simaan AbouRizk (University of Alberta) for alerting us to his dissertation as a source of the civil engineering data utilized in Chapters 1 and 4, and Dr. David Findley (U.S. Bureau of Census) for assisting us to obtain recent income distribution data for Chapter 7. Our thanks go to our student Caner Sener for careful proofreading of the manuscript and drawing our attention to a number of misprints. Needless to say, the responsibility for remaining misprints and errors is ours alone. The positive attitude and encouragement from the editors of the World Scientific Publications was very valuable. We are especially indebted to our editor Ji Zhang who generously devoted her time and have shown enthusiasm for the project. Ji Zhang is a marvelous editor with an uncanny sense on how to reshape a narrative; working with her was a delight. Finally, we want to express gratitude to our wives Greer and Rosalie for encouragement and displaying patience while we were toiling away obsessively with the manuscript and for providing a balance to our lives.

We trust that this monograph will fill a gap in the literature on statistical distributions and readers will find our efforts useful in their work. We welcome comments, suggestions and criticism from the readers.

> *Johan René van Dorp and Samuel Kotz*
> *September, 2004*
> *Washington, DC, U.S.A.*

Contents

Chapter 1

The Triangular Distribution

One of our goals in this book is to "dig out" suitable substitutes of the beta distribution. Only recently (less than 10 years ago) has the triangular distribution *specifically* been investigated by D. Johnson (1997) as a *proxy for the beta distribution*, even though its origins can be traced back to Thomas Simpson (1755) (about one century after the discovery of the beta distribution in a letter from Sir Isaac Newton to Henry Oldenberg). Very recently a "Handbook of Beta Distributions" edited by Gupta and Nadarajah (2004) has appeared (providing and emphasizing in a single monograph the attention that the beta distribution has attracted by both statistical theoreticians and practitioners over the last century, or so). On the other hand it appears that, in our opinion, the triangular distribution has been somewhat neglected in the statistical literature (perhaps even due to its simplicity which may discourage research efforts). In this chapter, we shall attempt to provide some chronology regarding the history of this distribution, state some of its properties and describe methods for estimating its parameters. Although the exposition is certainly not complete, we hope that it becomes apparent that the triangular distributions' "simplicity" is to a certain extent wrongly perceived and these distributions and their extensions are certainly worthy of further investigations.

1.1 An Historical Overview

Written records on the triangular distribution seem to originate in the middle of the 18-th century when problems of combinatorial probability were at their peak. A historically inclined reader may wish to consult the classical book by F.N. David (1962). One of the earliest mentions of the triangular distributions seems to be in Simpson[1] (1755, 1757). Thomas Simpson was a colorful personality in Georgian England. His life and adventures are described — in somewhat unflattering terms — in Pearson (1978) and Hald (1990). (Stigler (1986) gives a more sympathetic assessment of Simpson's work and character.) Stigler (1984) and more recently Farebrother (1990) provide some additional details on Thomas Simpson in particular on the correspondence with Roger Boscovich (1711-1787) a famous Italian astronomer and statistician of Serbian origin. The correspondence deals with the method of least absolute deviations regression problem which indirectly relates to triangular distributions (see, Farebrother (1990) and Stigler (1984)).

According to Seal (1949), Simpsons' object was to consider mathematically the method 'practised by Astronomers' of taking the mean of several observational readings "in order to diminish the errors arising from the imperfection of instruments and of the organs of sense". He supposes that any one reading errors in excess or defects are symmetrically disposed and have assignable upper and lower limits. He gives the probability that the mean of n observations falls between the boundaries $\pm z$ for the following discrete asymmetric triangular probability law:

$$p(x) = \begin{cases} \frac{h+1-|x|}{(h+1)^2} & x = -h, -h+1, \ldots -1, 0, 1, \ldots, h. \\ 0 & x = \text{any other value.} \end{cases} \tag{1.1}$$

The solution for the case of a uniform discrete distribution, expressed as a gaming problem via a generalized die with k faces, was known by 1710, and Simpson's treatment by means of generating functions is the same as Abraham de Moivre[2] (1667-1754) (Todhunter (1865), p.85, Hald (1990)) which caused accusations of plagiarism. What is novel in Simpson's work appears in the four pages of additional material published in 1757. Here he extends the solution for the triangular case (1.1) to the limiting case $h \to \infty$ in such a way that the range of variation of an individual error remains

---

[1]Thomas Simpson (1710-1761) a prolific writer of mathematical textbooks and able teacher at the Royal Military Acadamy in Wolwich England has made original and important contributions to actuarial sciences.

[2]Abraham De Moivre (from a Huguenot family) left France in 1685 to seek asylum in England. He was a promininent probabilist who was the first to provide the normal approximation to the binomial distribution.

within ± 1. Seal (1949) points out that this is the *first* time a continuous (symmetric triangular) probability law is introduced. Hence, the continuous triangular distribution is certainly amongst the first *continuous* distributions to have been noticed by investigators during the 18-th century (when these types of problems were popular). For example, one of the first records that mentions the *continuous* uniform distribution is the famous paper by the reverend Thomas Bayes (1763) (only a few years after Simpons' written records in 1757).

The symmetric triangular distribution with probability density function (pdf)

$$f(x) = \begin{cases} 4x, & \text{for } 0 \leq x \leq \frac{1}{2}, \\ 4(1-x), & \text{for } \frac{1}{2} \leq x \leq 0, \\ 0, & \text{elsewhere} \end{cases} \qquad (1.2)$$

and support $[0,1]$ is depicted in Fig. 1.1A. R. Schmidt (1934) possibly was the first to notice that the pdf (1.2) follows as the distribution of the arithmetic average of two uniform random variables U_1 and U_2 on $[0,1]$, i.e.

$$X = \frac{U_1 + U_2}{2}. \qquad (1.3).$$

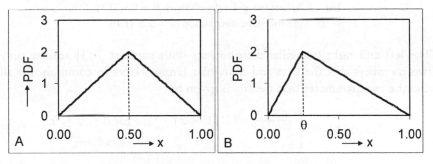

Fig. 1.1 A: Standard symmetric triangular distribution
B: Standard asymmetric triangular distribution with $\theta = 1/4$.

He referred to it as a *tine distribution* ("tine" is a slender projecting point). We were not able to find other Western sources dealing with triangular distributions between Simpson (1757) and Schmidt (1934) in the mainstream statistical literature. Asymmetric standard triangular distributions support $[0, 1]$ were studied by Ayyangar (1941). The pdf is given by

$$f(x|\theta) = \begin{cases} 2\frac{x}{\theta}, & \text{for } 0 \le x \le \theta, \\ 2\frac{1-x}{1-\theta}, & \text{for } \theta \le x \le 1, \\ 0, & \text{elsewhere.} \end{cases} \qquad (1.4)$$

Substituting $\theta = 1/2$ yields the pdf (1.2). A standard asymmetric triangular distribution with $\theta = 1/4$ is depicted in Fig. 1.1B. The left ($\theta = 0$) and right ($\theta = 1$) triangular distribution (discussed in Rider (1963)) are depicted in Figs. 1.2A and 1.2B, respectively.

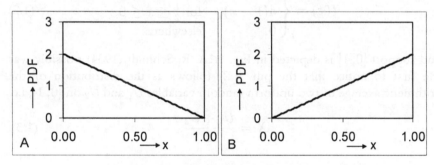

Fig. 1.2 A: Left triangular distribution ($\theta = 0$ in (1.4));
B: Right triangular distribution ($\theta = 1$ in (1.4)).

The left and right triangular distributions with support $[0, 1]$ are the only two members that the beta and triangular families have in common. Recall that the two parameter beta density is given by

$$f(x|\alpha, \beta) = \begin{cases} \frac{\Gamma(\alpha+\beta)}{\Gamma(\alpha)\Gamma(\beta)}x^{\alpha-1}(1-x)^{\beta-1}, & \text{for } 0 \le x \le 1 \\ 0, & \text{elsewhere,} \end{cases} \qquad (1.5)$$

where $\alpha > 0, \beta > 0$ and $\Gamma(\,\cdot\,)$ is the gamma function. Substituting $\alpha = 1$ and $\beta = 2$ in the beta pdf (1.5) yields the left triangular pdf ($\theta = 0$ in (1.4)). Substituting $\alpha = 2$ and $\beta = 1$ yields the right triangular one ($\theta = 1$ in (1.4)). Since 1941 up to the mid-sixties very few publications were devoted to the triangular distribution (Fullman (1953), Ostle *et al.* (1961) and Rider (1963)). The product of two identically independent distributed (i.i.d.) triangular random variables has been investigated by Donahue (1964). The sum of two independent triangular random variables (i.e., their convolution) sharing the same support (but not necessarily with the same mode) has — to the best of our knowledge — only been investigated very recently by Van Dorp and Kotz (2003b).

Since 1962 up to 1999, the distribution emerges in numerous papers dealing with the Project Evaluation and Review Technique — PERT (see, e.g., Clark (1962), Grubbs (1962), MacCrimmon and Ryaveck (1964), Moder and Rodgers (1968), Vāduva (1971), Williams (1992), Keefer and Verdini (1993), and D. Johnson (1997) amongst others). These papers deal with the asymmetric three-parameter triangular density

$$f(z|a,m,b) = \begin{cases} \frac{2}{b-a}\frac{z-a}{m-a}, & \text{for } a \leq z \leq m, \\ \frac{2}{b-a}\frac{b-z}{b-m}, & \text{for } m \leq z \leq b, \\ 0, & \text{elsewhere} \end{cases} \qquad (1.6)$$

(with support $[a,b]$ and the mode m) which by means of the transformation $z = (x-a)/(b-a)$ reduces to its standard form (1.4) with the support $[0,1]$, where $\theta = (m-a)/(b-a)$. The parameters of the triangular distribution (1.6) are in one-to-one correspondence with a lower estimate $\widehat{a}$, a most likely estimate $\widehat{m}$, and an upper estimate $\widehat{b}$ of a characteristic under consideration. This leads to an intuitive appeal of the triangular distribution (see, e.g., Williams (1992)). In PERT these characteristics are the completion times of activities in a project network (see, Winston (1993)) whose uncertainties may be modeled by the distribution (1.6). N.L. Johnson and Kotz (1999) discuss the asymmetric triangular distribution in the context of YAWL distributions which have *inter alia* applications in modeling prices associated with orders placed by investors for single securities traded on the New York and American Stock Exchanges.

Recent popularity of the triangular distribution can be attributed to its use in Monte Carlo simulation modeling (see, e.g., Vose (1996) and Garvey (2002)), discrete system simulation (see, e.g., Banks *et al.* (2000), Altiok and Melamed (2001), Kelton *et al.* (2002)) and its use in standard uncertainty analysis software — such as @Risk (developed by the Palisade Corporation) or Crystal Ball (developed by Decision Engineering). These books and packages recommend the use of the triangular distribution when the underlying distribution is unknown, but a minimal value $\widehat{a}$, some maximal value $\widehat{b}$ and a most likely value $\widehat{m}$ are available. In Chapter 4, we shall discuss in some detail the appropriateness of this modeling approach given *only* these estimates.

1.2 Deriving the CDF utilizing a Geometric Argument

Instead of deriving the three-parameter cumulative distribution function (cdf) of the triangular distribution in the usual fashion from its pdf (1.6), we shall derive it using a geometric argument involving triangles (from which the triangular distribution derives its name). Figure 1.3A depicts the density function of a triangular distribution with parameters a, m and b, splitting the area underneath it into two triangles with area A_1 and A_2, respectively.

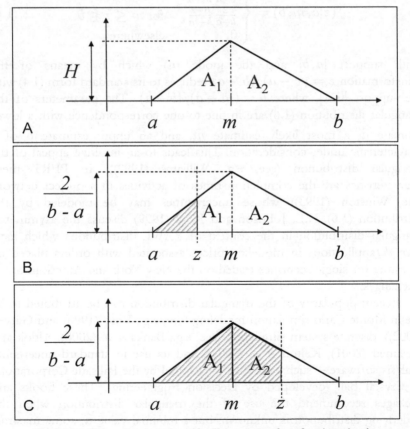

Fig. 1.3 Deriving of a triangular cdf utilizing areas of conforming triangles.

Since, from basic properties of a pdf it follows that $A_1 + A_2 = 1$ we have (see Fig. 1.3A)

$$(m - a)\frac{H}{2} + (b - m)\frac{H}{2} = 1 \Leftrightarrow H = \frac{2}{b - a}. \tag{1.7}$$

Hence, the density value at the mode m is not a function of the location of m, relative to the boundaries a and b (which is not obvious). Note that in Figs. 1.1 and 1.2 the density value at the mode equals 2 in all cases (since $a = 0$ and $b = 1$). In addition, from (1.7) we have

$$A_1 = \frac{m - a}{b - a} \text{ and } A_2 = \frac{b - m}{b - a}. \tag{1.8}$$

In other words, the probability mass to the left (the right) of the mode m, equals the *relative* distance of the mode m to the lower bound a (the upper bound b) compared to the whole range from a to b.

From Fig. 1.3B, Eq. (1.8) and utilizing conformity of the triangles, it immediately follows that for $a \leq z \leq m$:

$$Pr(Z \leq z) = \left(\frac{z - a}{m - a}\right)^2 A_1 = \frac{m - a}{b - a}\left(\frac{z - a}{m - a}\right)^2 \tag{1.9}$$

and for $m \leq z \leq b$ using Fig. 1.3C and the complement rule $Pr(Z \leq z) = 1 - Pr(Z > z)$:

$$Pr(Z \leq z) = 1 - \left(\frac{b - z}{b - m}\right)^2 A_2 = 1 - \frac{b - m}{b - a}\left(\frac{b - z}{b - m}\right)^2 \tag{1.10}$$

Hence, the cdf is given by:

$$F(z) = Pr(Z \leq z) = \begin{cases} \frac{m-a}{b-a}\left(\frac{z-a}{z-a}\right)^2, & \text{for } a \leq z \leq m, \\ 1 - \frac{b-m}{b-a}\left(\frac{b-z}{b-m}\right)^2 & \text{for } m \leq z \leq b. \end{cases} \tag{1.11}$$

Taking the derivative with respect to z in (1.11) we arrive at the pdf (1.6). The reader may wish to graph the cdf (1.11) for reasonable choices of a, m and b ($a \leq m \leq b$).

The inverse cdf of Z follows from (1.11) as

$$F^{-1}(y|a, m, b, n) = \quad (1.12)$$

$$\begin{cases} a + \sqrt{y(m-a)(b-a)}, & \text{for } 0 \leq y \leq \frac{m-a}{b-a} \\ b - \sqrt{(1-y)(b-m)(b-a)}, & \text{for } \frac{m-a}{b-a} \leq y \leq 1. \end{cases}$$

Equation (1.12) allows for straightforward sampling from a triangular distribution with support $[a, b]$ utilizing the inverse cdf transformation technique and a pseudo-random number generator of a uniform random variable on $[0, 1]$ (see, e.g., Vose (1996)). Pseudo random number generators have become standard in spreadsheet software and are also utilized in uncertainty analysis packages such as @Risk (developed by the Palisade Corporation) and Crystal Ball (developed by Decision Engineering), and discrete event simulation software such as Arena (developed by Rockwell Software). The quality of the sample from a triangular distribution utilizing the inverse cdf transformation technique is identical to that obtained using the pseudo-random number generator. Banks *et al.* (2000) provide an excellent overview of desirable properties of and statistical tests for uniformity and independence of pseudo random number generators.

1.3 Moments of Triangular Distributions

The k-th moment about zero (which we shall denote by μ'_k) of a standard triangular distribution with support $[0, 1]$ follows from the pdf (1.4) as

$$\mu'_k = E[X^k] = \int_0^1 x^k f(x|\theta)dx = \frac{2(1 - \theta^{k+1})}{(k+1)(k+2)(1-\theta)}. \quad (1.13)$$

Here calculations are a bit lengthy but straightforward. The corresponding moments of a triangular variable Z with support $[a, b]$ and pdf (1.6) follow from (1.13) and the linear transformation $Z = (b - a)X + a$. Specifically,

$$E[Z^k] = E[\{(b-a)X + a\}^k] = \sum_{i=0}^{k} \binom{k}{i}(b-a)^i a^{k-1}E[X^i]. \quad (1.14)$$

Substituting $k = 1$ and $k = 2$ in (1.13) we arrive at the first and the second moments about zero of X:

$$\mu_1' = E[X] = \frac{\theta + 1}{3} \; ; \mu_2' = E[X^2] = \frac{(1 - \theta^3)}{6(1 - \theta)} \tag{1.15}$$

and from the relation $\mu_2 = Var(X) = E[X^2] - E^2[X]$ we have

$$Var(X) = \mu_2 = \frac{1 - \theta(1 - \theta)}{18}. \tag{1.16}$$

Hence, the variance attains its minimum $3/72$ at $\theta = 1/2$ and its maximum $1/18$ at $\theta = 0$ or $\theta = 1$. Recall that the variance of a standard uniform distribution is much larger and equal to $1/12$.

In a similar manner, utilizing (1.15), substituting $k = 3$ and $k = 4$ in (1.13) and applying the definitions of the central moments

$$\begin{cases} \mu_3 = E[(X - E[X])]^3 = \mu_3' - 3\mu_2'\mu_1' + 2\mu_1'^3 \\ \mu_4 = E[(X - E[X])]^4 = \mu_4' - 4\mu_3'\mu_1' + 6\mu_2'\mu_1'^2 - 3\mu_1'^4 \end{cases} \tag{1.17}$$

one obtains (see, Johnson and Kotz (1999)):

$$\begin{cases} \mu_3 = \frac{1}{270}(1 - 2\theta)(2 - \theta)(1 - \theta) \\ \mu_4 = \frac{1}{135}\{1 - \theta(1 - \theta)\}^2 \end{cases}$$

From the definitions of skewness $\sqrt{\beta_1}$ and kurtosis β_2 (see, e.g., Stuart and Ord (1994)) :

$$\sqrt{\beta_1} = Sign\{\mu_3\}\sqrt{\frac{\mu_3^2}{\mu_2^3}}; \; \beta_2 = \frac{\mu_4}{\mu_2^2}, \tag{1.18}$$

(the skewness $\sqrt{\beta_1}$ retains the sign of the third central moment μ_3) we have

$$\sqrt{\beta_1} = \frac{(1 - 2\theta)(2 - \theta)(1 - \theta)}{\{1 - \theta(1 - \theta)\}^{\frac{3}{2}}} \frac{1}{5}\sqrt{2} \tag{1.19}$$

and

$$\beta_2 = 2.4. \tag{1.20}$$

(Compare with the kurtosis of a normal or Gaussian distribution, which equals 3.) The kurtosis β_2 (which is a combined measure of peakedness and heaviness of the tails of a distribution) here does not depend on θ.

Figure 1.4 plots skewness $\sqrt{\beta_1}$ as a function of θ. Observe that minimum skewness $-\frac{2}{5}\sqrt{2} \approx -0.566$ (which is a negative value) is attained for the right triangular distribution in Fig. 1.2B. It is important to note here that the left skewed distribution (with a heavier tail towards the left) has negative skewness and thus the designation *right* triangular distribution in Fig. 1.2B arises from the location of the mode θ being at the right boundary of the support. Similarly, the right skewed, *left* triangular distribution in Fig. 1.2A has the maximum positive skewness $\frac{2}{5}\sqrt{2} \approx 0.566$. The skewness of a symmetric triangular distributions $\sqrt{\beta_1} = 0$ is obtained from (1.19) by substituting $\theta = \frac{1}{2}$ (see Fig. 1.4).

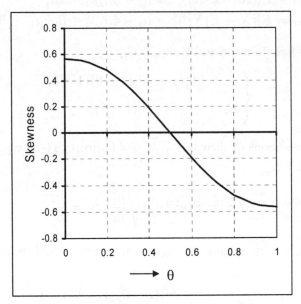

Fig. 1.4 Skewness $\sqrt{\beta_1}$ (Eq. (1.19)) as a function of θ.

Since the measures skewness and kurtosis are invariant under linear scale transformation it follows that (1.19) and (1.20), respectively, may be used for a triangular random variable Z with support $[a, b]$, pdf (1.6) and parameters a, m and b, utilizing $\theta = (m - a)/(b - a)$. From the linear transformation $Z = (b - a)X + a$, (1.15) and (1.16) we derive

$$E[Z] = \frac{a + m + b}{3} \qquad (1.21)$$

and

$$Var[Z] = \frac{(b - a)^2}{18} \cdot \left\{ 1 - \frac{m - a}{b - a} \frac{b - m}{b - a} \right\}. \qquad (1.22)$$

Note that from (1.21) it follows that the mean value of Z is the arithmetic average of the lower bound a, the mode m and the upper bound b. In our opinion, the popularity of the triangular distribution arises from the straightforward relationship (1.21) between the parameters and the mean of Z, a meaningful interpretation of the parameters a, m and b as well as from the property that the probability mass to the left of the mode m equals the relative distance of the mode m to the lower bound a over the whole support $[a, b]$ (i.e. $(m - a)/(b - a)$, see Eq. (1.8)).

1.4 Maximum Likelihood Method for the Threshold Parameter θ.

The structure of the standard triangular distribution (1.4) with support $[0, 1]$ leads to an illuminating procedure for estimating the threshold parameter θ. This parameter can be viewed as "dividing" (in the sense that it is related to two different analytical expressions appearing in the definition of the pdf (1.4)). The derivation of the ML estimator for θ in (1.4) seems to be quite instructive (and is similar, but simplified compared to the one presented in Johnson and Kotz (1999)).

Let for a random i.i.d. sample of size s, $\underline{X} = (X_1, \ldots, X_s)$, the order statistics be $X_{(1)} < X_{(2)} \ldots < X_{(s)}$. By definition, the likelihood for X with distribution (1.4) is

$$L(\underline{X} ; \theta) = 2^s \{H(\underline{X}; \theta)\} \qquad (1.23)$$

where

$$H(\underline{X};\theta) = \frac{\prod\limits_{i=1}^{r} X_{(i)} \prod\limits_{i=r+1}^{s} (1 - X_{(i)})}{\theta^r (1 - \theta)^{s-r}} \tag{1.24}$$

and r is implicitly defined by $X_{(r)} \le \theta < X_{(r+1)}$, $X_{(0)} \equiv 0$ and $X_{(s+1)} \equiv 1$.

Theorem 1.1: Let $\underline{X} = (X_1, \ldots, X_s)$ be an i.i.d. sample from a triangular distribution with the pdf (1.4) and support $[0, 1]$. The ML estimator of θ maximizing the likelihood (1.23) over the parameter domain $0 \le \theta \le 1$ is

$$\widehat{\theta} = X_{(\widehat{r})}, \tag{1.25}$$

where

$$\widehat{r} = \underset{r \in \{1, \ldots, s\}}{\arg\max} M(r) \tag{1.26}$$

and

$$M(r) = \prod_{i=1}^{r-1} \frac{X_{(i)}}{X_{(r)}} \prod_{i=r+1}^{s} \frac{1 - X_{(i)}}{1 - X_{(r)}}. \tag{1.27}$$

Proof: We shall provide a detailed proof of this basic theorem. (Another version of this theorem will be encountered in Chapter 5). To maximize the likelihood (1.23), we represent it as

$$\underset{0 \le \theta \le 1}{\max} L(\underline{X};\theta) = 2^s \widehat{M}, \tag{1.28}$$

where

$$\widehat{M} = \underset{0 \le \theta \le 1}{\max} H(\underline{X};\theta), \tag{1.29}$$

$H(\underline{X};\theta)$ is defined by (1.24) and $X_{(r)} \le \theta \le X_{(r+1)}$, with $X_{(0)} \equiv 0$, $X_{(s+1)} \equiv 1$. Utilizing (1.29) one can therefore write

$$\widehat{M} = \underset{r \in \{0, \ldots, s\}}{\max} H(r), \tag{1.30}$$

where

$$H(r) = \max_{X_{(r)} \le \theta \le X_{(r+1)}} H(\underline{X}; \theta), \tag{1.31}$$

$r = 0, \ldots, s$, $X_{(0)} \equiv 0$ and $X_{(s+1)} \equiv 1$. The three non-overlapping cases: $r \in \{1, \ldots, s-1\}$, $r = 0$ and $r = s$ will be discussed separately.

Case $r \in \{1, \ldots, s-1\}$: Here, $X_{(r)} \le \theta \le X_{(r+1)}$. The function

$$g(\theta) = \theta^r (1 - \theta)^{s-r} \tag{1.32}$$

in the denominator of the definition of $H(\underline{X}; \theta)$ (1.24) is proportional to an unimodal beta density since $r \in \{1, \ldots, s-1\}$. Thus,

$$\min_{X_{(r)} \le \theta \le X_{(r+1)}} g(\theta) = \min_{\theta \in \{X_{(r)}, X_{(r+1)}\}} g(\theta) \tag{1.33}$$

and, from (1.24), (1.31) and (1.33),

$$H(r) = \max_{r' \in \{r, r+1\}} \prod_{i=1}^{r'-1} \frac{X_{(i)}}{X_{(r')}} \prod_{i=r'+1}^{s} \frac{1 - X_{(i)}}{1 - X_{(r')}}. \tag{1.34}$$

Case $r = 0$: Here $0 \le \theta \le X_{(1)}$. From (1.24) and (1.31) it follows that now

$$H(0) = \max_{0 \le \theta \le X_{(1)}} \prod_{i=1}^{s} \frac{1 - X_{(i)}}{1 - \theta}.$$

Hence $H(0)$ becomes the product

$$H(0) = \prod_{i=1}^{s} \frac{1 - X_{(i)}}{1 - X_{(1)}} \equiv \prod_{i=2}^{s} \frac{1 - X_{(i)}}{1 - X_{(1)}}. \tag{1.35}$$

Case $r = s$: Here $X_{(s)} \le \theta \le 1$. From (1.24) and (1.31) it follows that in this case

$$H(s) = \max_{X_{(s)} \le \theta \le 1} \prod_{i=1}^{s} \frac{X_{(i)}}{\theta}.$$

Hence $H(s)$ becomes the product

$$H(s) = \prod_{i=1}^{s} \frac{X_{(i)}}{X_{(s)}} \equiv \prod_{i=1}^{s-1} \frac{X_{(i)}}{X_{(s)}}. \tag{1.36}$$

(Compare with (1.35).) From (1.30), (1.34), (1.35) and (1.36) we obtain that

$$\widehat{M} = \max_{r \in \{1, \ldots, s\}} M(r), \tag{1.37}$$

where $M(r)$ is defined by (1.27). Hence, the ML estimator of the threshold parameter θ equals the order statistic $X_{(\widehat{r})}$, where $\widehat{r}$ is given by (1.26). $\square$

The ML estimator $\widehat{\theta} = X_{(\widehat{r})}$ given in (1.25) is quite intuitive (if one recalls the ML estimator $\widehat{\theta} = X_{(s)}$ of the parameter of a uniform distribution on $[0, \theta]$ for a sample of size s).

1.4.1 An illustrative example

We shall illustrate the ML estimation procedure for the parameter θ of a standard triangular distribution (1.4) by means of the following hypothetical order statistics

$$(X_{(1)}, \ldots, X_{(8)}) = (0.10, 0.25, 0.30, 0.40, 0.45, 0.60, 0.75, 0.80). \tag{1.38}$$

This data was also used in Johnson and Kotz (1999)[3]. Consider the matrix $A = [a_{i,r}]$ with the entries :

$$a_{i,r} = \begin{cases} \frac{X_{(i)}}{X_{(r)}} & i < r \\ \frac{1-X_{(i)}}{1-X_{(r)}} & i \geq r. \end{cases} \tag{1.39}$$

Table 1.1 summarizes calculations of the matrix A for the order statistics given in (1.38). The last row in the table contains the products of the matrix entries in the r-th column which are equal to the values of $M(r)$ given by

---

[3]Note: The values in Johnson and Kotz (1999) corresponding to the last three entries in the last row of Table 1.1 contain the following typos; 0.00547 should read 0.00543, 0.00137 should replace 0.00364 and 0.00029 should be 0.00290.

(1.27), $r = 1, \ldots, s$. Here $s = 8$. From the last row of Table 1.1 and utilizing (1.37), (1.26) and (1.25) we calculate

$$\widehat{M} = 0.011; \; \widehat{r} = 3;$$
$$\widehat{\theta} = X_{(\widehat{r})} = 0.30. \tag{1.40}$$

Table 1.1 ML estimation for a triangular distribution
with pdf (1.4) using the data given by (1.38).

r		1	2	3	4	5	6	7	8
		$X_{(1)}$	$X_{(2)}$	$X_{(3)}$	$X_{(4)}$	$X_{(5)}$	$X_{(6)}$	$X_{(7)}$	$X_{(8)}$
i		0.10	0.25	0.30	0.40	0.45	0.60	0.75	0.80
1	$X_{(1)}$ 0.10	1	0.400	0.333	0.250	0.222	0.167	0.133	0.125
2	$X_{(2)}$ 0.25	0.833	1	0.833	0.625	0.556	0.417	0.333	0.313
3	$X_{(3)}$ 0.30	0.778	0.933	1	0.750	0.667	0.500	0.400	0.375
4	$X_{(4)}$ 0.40	0.667	0.800	0.857	1	0.889	0.667	0.533	0.500
5	$X_{(5)}$ 0.45	0.611	0.733	0.786	0.917	1	0.750	0.600	0.563
6	$X_{(6)}$ 0.60	0.444	0.533	0.571	0.667	0.727	1	0.800	0.750
7	$X_{(7)}$ 0.75	0.278	0.333	0.357	0.417	0.455	0.625	1	0.938
8	$X_{(8)}$ 0.80	0.222	0.267	0.286	0.333	0.364	0.500	0.800	1
M(r)		0.007	0.010	*0.011*	0.010	0.009	0.005	0.004	0.003

Figure 1.5 displays the function $H(\underline{X}; \theta)$ defined by Eq. (1.24) and shows that for the data in (1.38) the maximum value $\widehat{M} = 0.011$ of $H(\underline{X}; \theta)$ over $\theta \in [0, 1]$ is attained at $X_{(3)} = 0.30$. From (1.28), $\widehat{M} = 0.011$ and $s = 8$ we have $L(\underline{X}; \widehat{\theta}) \approx 2.79$. Also observe that the maximum value $H(r)$ (see, Eq. (1.31)) of $H(\underline{X}; \theta)$ over $\theta \in [X_{(r)}, X_{(r+1)}]$ is attained at either $X_{(r)}$ or $X_{(r+1)}$ for all $r = 0, \ldots, s$.

The ML estimation of the mode θ of the triangular pdf (1.4) with support $[0, 1]$ can easily be modified to the ML estimation of the mode m of the triangular pdf (1.6) with support $[a, b]$, using the linear scale transformation $Z = (b - a)X + a$; recall that the parameters a and b are fixed and the parameter $m = (b - a)\theta + a$. The ML estimator of the parameter m of the distribution (1.6) utilizing the order statistics $(Z_{(1)}, \ldots, Z_{(s)})$ are

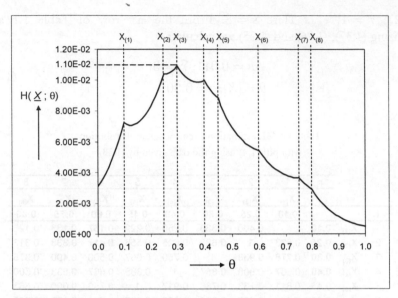

Fig. 1.5 Graph of $H(X;\theta)$ (1.24) for the data in (1.38). (Observe that maxima over the sets $[X_{(r)}, X_{(r+1)}]$ are attained here solely at the order statistics, $r = 1, \dots, s$).

$$\widehat{m}(a,b) = Z_{(\widehat{r}(a,b))} \tag{1.41}$$

where, as above,

$$\widehat{r}(a,b) = \underset{r \in \{1, \dots, s\}}{arg\ max}\ M(a,b,r) \tag{1.42}$$

and

$$M(a,b,r) = \prod_{i=1}^{r-1} \frac{Z_{(i)} - a}{Z_{(r)} - a} \prod_{i=r+1}^{s} \frac{b - Z_{(i)}}{b - Z_{(r)}}. \tag{1.43}$$

Compare with equations (1.25), (1.26) and (1.27).

1.5 Three Parameter Maximum Likelihood Estimation

This lengthy section involves some non-standard interesting derivations of the ML procedure of the three-parameter triangular distributions which are

16

closely related to a non-regular case of ML estimation for continuous distributions (see, e.g., Cheng and Amin (1983)).

Let Z be a random variable with pdf (1.6). For a random sample $\underline{Z} = (Z_1, \ldots, Z_s)$ with size s from a triangular distribution with support $[a, b]$ and mode m, let the order statistics be $Z_{(1)} < Z_{(2)} < \ldots < Z_{(s)}$. Utilizing (1.6), the likelihood for $\underline{Z}$ is by definition

$$L(\underline{Z}; a, m, b) = \left(\frac{2}{b-a}\right)^s \left\{ \prod_{i=1}^r \frac{Z_{(i)} - a}{m - a} \prod_{i=r+1}^s \frac{b - Z_{(i)}}{b - m} \right\} \quad (1.44)$$

where r is implicitly defined by $Z_{(r)} \leq m < Z_{(r+1)}$, $Z_{(0)} \equiv a$ and $Z_{(s+1)} \equiv b$. Thus, analogously to (1.28) it follows that for fixed values of a and b, satisfying

$$a < Z_{(1)} \text{ and } b > Z_{(s)},$$

we have

$$\max_{a \leq m \leq b} L(\underline{Z}; a, m, b) = \left(\frac{2}{b-a}\right)^s \left\{ M(a, b, \widehat{r}(a, b)) \right\} \quad (1.45)$$

where $\widehat{r}(a, b)$ and $M(a, b, r)$ are given by (1.42) and (1.43), respectively. The ML estimator for the mode m (as a function of a and b) is given by Eq. (1.41). Note that, the function $\widehat{r}(a, b)$ is an *index* function indicating at which order statistic the ML estimate of the parameter m is attained as a function of the lower bound a and upper bound b (we shall elaborate on the index function $\widehat{r}(a, b)$ below).

From (1.45) we have that

$$\max_{S(a, m, b)} \left[Log\{L(\underline{Z}; a, m, b)\} \right] = \quad (1.46)$$

$$\max_{a < X_{(1)}, b > X_{(s)}} \left[sLog2 + G(a, b) \right],$$

where the set

$$S(a, m, b) = \quad (1.47)$$
$$\{(a, m, b) \mid a < Z_{(1)}, b > Z_{(s)}, a \leq m \leq b\}$$

and the function

$$G(a, b) = Log\{M(a, b, \widehat{r}(a, b))\} - sLog\{b - a\}. \quad (1.48)$$

This is an interesting function to be discussed below. Again recall the definitions of $M(a, b, r)$ and that of $\widehat{r}(a, b)$ in Eqs. (1.42) and (1.43). Note that $G(a, b)$ is only defined for values of $a < Z_{(1)}$ and $b > Z_{(s)}$ (see Eq. (1.47)). To summarize, the three-dimensional optimization problem of maximizing the likelihood (1.44) reduces to a two-dimensional case of maximizing $G(a, b)$ over the region $a < Z_{(1)}$ and $b > Z_{(s)}$. From the structure of (1.44), however, we can immediately conclude that for all values of m such that

$$Z_{(1)} < m < Z_{(s)} \quad (1.49)$$

the likelihood $L(\underline{Z}; a, m, b) \to 0$ (and hence $Log\{L(\underline{Z}; a, m, b)\} \to -\infty$) when $a \uparrow Z_{(1)}$ or $b \downarrow Z_{(s)}$. Thus, when a modal value can be observed in the data (via, for example, a histogram) indicating the validity of Eq. (1.49), it would seem that the ML estimators for a and b are *not* the order statistics $Z_{(1)}$ and $Z_{(s)}$, respectively. This is in contrast with the well-known fact that the ML estimators of a uniform distribution with support $[a, b]$ *are* given by smallest order statistic $X_{(1)}$ and the largest one $X_{(s)}$ (see, e.g., Devore (2004)).

We shall demonstrate the above fitting characteristic of a triangular distribution for civil engineering data consisting of a sample of 85 hauling times (Source, AbouRizk (1990)) rather than the hypothetical 8 point example given by (1.38) since we are now fitting a three parametric distribution instead of a distribution with one parameter θ given by (1.4). Figure 1.6 depicts the empirical pdf for the data in Table 1.2 which seems to have a mode in the vicinity of the center of the range $[Z_{(1)}, Z_{(85)}] = [3.20, 8.60]$. Hence, Eq. (1.49) is satisfied. In addition, Fig. 1.6 depicts the ML fitted triangular distribution with ML estimates of parameters

Table 1.2 Civil engineering data consisting of 85 hailing times
(Source: AbouRizk (1990)).

4.79	4.75	5.40	4.70	6.50
5.30	6.00	5.90	4.80	6.70
6.00	4.95	7.90	5.40	3.50
4.54	6.90	5.80	5.40	5.70
8.00	5.40	5.60	7.50	7.00
4.60	3.20	3.90	5.90	3.40
5.20	5.90	4.40	5.20	7.40
5.70	6.00	3.60	6.20	5.70
5.80	5.90	6.00	5.15	6.00
4.82	5.90	6.00	7.30	7.10
4.73	5.90	3.60	6.30	7.00
5.10	6.00	6.60	4.40	6.80
5.60	5.90	5.90	8.60	6.00
5.80	5.40	6.50	4.80	6.40
4.15	4.90	6.50	8.20	7.00
8.50	5.90	4.40	5.80	4.30
5.10	5.90	4.70	3.50	6.80

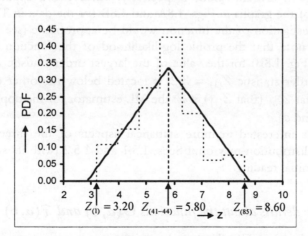

Fig. 1.6 Empirical pdf for the data in Table 1.2 together with a ML fitted three-parameter triangular distribution $\widehat{a} = 2.87$, $\widehat{m} = Z_{(41-44)} = 5.80$, $\widehat{b} = 8.80$.

$$\hat{a} = 2.87 < Z_{(1)} = 3.20, \ \hat{m} = Z_{(41-44)} = 5.80, \qquad (1.50)$$
$$\hat{b} = 8.80 > Z_{(s)} = 8.60.$$

Observe that the example data in Table 1.2 actually contains ties, resulting in the ML estimator $\hat{m}$ to be attained at either one of the order statistics $Z_{(41)}$ through $Z_{(44)}$. Also note that the triangular distribution in Fig. 1.6 does not quite capture the 'peak' of the empirical pdf in Fig. 1.6. In Chapter 4 we shall fit a four parameter generalization of the triangular distribution that does capture this 'peak' and present a more formal fit analysis using the chi-square test (see, e.g., Devore (2004)).

Figure 1.7 provides the form of the function $G(a,b)$ given by (1.48) that was maximized to arrive at the ML estimators for the lower and upper bounds a and b in (1.50) for the data in Table 1.2. Figure 1.8A (Figure 1.8B) depicts a likelihood profile of the function $G(a,b)$ displayed in Fig. 1.7 for the data in Table 1.2 as function of the parameter a (parameter b) for different fixed values of the parameter b (parameter a). Note the behavior of $G(a,b)$ for $b = 8.6$ (for $a = 3.2$) in Fig. 1.8A (Fig. 1.8B). The ML estimates $\hat{a} = 2.87$ and $\hat{b} = 8.80$ are indicated by means of a vertical solid line in Figs. 1.8A and 1.8B, respectively. Observe the apparent mirror symmetry of the graphs in Figs. 1.8A and 1.8B for the data in Table 1.2. A further investigation of the function would be appropriate (see Sec. 1.5.1). Moreover, note that the profile log-likelihood of the function $G(a,b)$ in Fig. 1.8A (Fig. 1.8B) for the value of the largest order statistic $Z_{(s)} = 8.6$ (smallest order statistic $Z_{(1)} = 3.2$) is located below the other two, which indicates that $Z_{(s)}$ (that $Z_{(1)}$) is *not* the ML estimator for the upper bound b (lower bound a).

Readers interested in more statistical aspects of the three-parameter triangular distribution may omit Secs. 1.5.1 and 1.5.2 (with its subsections) during an initial reading.

1.5.1 Some details about the functions $G(a,b)$ and $\hat{r}(a,b)$

While the function $G(a,b)$ given by (1.48) is continuous over its domain $a < Z_{(1)}$ and $b > Z_{(s)}$, the partial derivatives with respect to a or b may not be unique at a finite $(s - 1)$ number of points. The source of non-differentiability at these points is due to the behavior of the index function

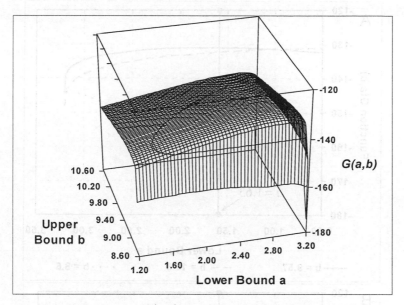

Fig. 1.7 The function $G(a, b)$ given by (1.48) for the data in Table 1.2.

$\widehat{r}(a, b)$ given by (1.42) as a function of the parameters a and b. In fact, the following properties can be derived for $\widehat{r}(a, b)$ as a function of b, keeping $a < X_{(1)}$ fixed (recall that $\widehat{r}(a, b)$ is an *index* function indicating at which order statistic the ML estimate of the parameter m is attained);

(1) The order statistic index $\widehat{r}(a, b)$ is decreasing in b;

(2) $\lim_{b \to \infty} \widehat{r}(a, b) = 1$;

(3) $\lim_{b \downarrow X_{(s)}} \widehat{r}(a, b) = s$;

(4) $\widehat{r}(a, b)$ as a function of b has $(s - 1)$ discontinuities at the points

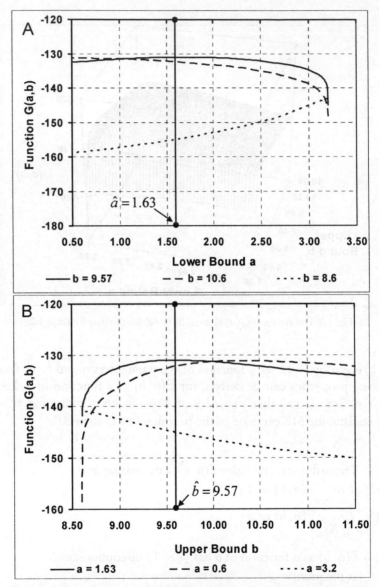

Fig. 1.8 Profiles of the function $G(a, b)$ given by (1.48) for the data in Table 1.2:
Graph A: as a function of the lower bound a; Graph B: as a function of the upper bound b.
The ML estimates $\widehat{a} = 2.87; \widehat{b} = 8.80$ in (1.50) are indicated by means
of a vertical dotted line in Figs. 1.8A and 1.8B, respectively.

22

$$f_b(a,r) = \tag{1.51}$$

$$\frac{X_{(r+1)} - X_{(r)} \sqrt[s-r]{\left(\dfrac{X_{(r)} - a}{X_{(r+1)} - a}\right)^r}}{1 - \sqrt[s-r]{\left(\dfrac{X_{(r)} - a}{X_{(r+1)} - a}\right)^r}}, \quad r \in \{1, \dots, s-1\}.$$

(Note that the parameter a is fixed.) Similar properties can be derived for $\widehat{r}(a,b)$ as a function of a, while keeping $b > X_{(s)}$ fixed. Figure 1.9 gives the form of the function $\widehat{r}(a,b)$ (Eq. (1.42)) for the data in (1.38). The function $\widehat{r}(a,b)$ may be viewed as a bivariate step-function or a *winding staircase function*, which could serve as a useful tool for studying non-differentiable bivariate distributions. We are purposely using only a the small set of 8 data points in (1.38) in Fig. 1.9 to emphasize the stepwise behavior of the function $\widehat{r}(a,b)$, which would have been less apparent visually when using, for example, the whole data set in Table 1.2. The central axis of the "winding staircase" in Fig. 1.9 is located at $a = Z_{(1)}$ $= 0.10$ and $b = Z_{(s)} = 0.80$. For a fixed a, the value of $f_b(a,r)$ (1.51) identifies the location of the r-th step (in terms of b) of the winding stare case. Note that at the central axis $(a = X_{(1)}, b = X_{(s)})$, the $(s-1)$ discontinuities $f_b(a,r)$ of the index function $\widehat{r}(a,b)$ converge.

Discarding the points of discontinuity of the function $\widehat{r}(a,b)$, the function $G(a,b)$ becomes differentiable with respect to a and b. From (1.48) we obtain:

$$\frac{\partial}{\partial a}G(a,b) = \frac{\frac{\partial}{\partial a}M(a,b,\widehat{r}(a,b))}{M(a,b,\widehat{r}(a,b))} + \frac{s}{b-a} \tag{1.52}$$

and

$$\frac{\partial}{\partial b}G(a,b) = \frac{\frac{\partial}{\partial b}M(a,b,\widehat{r}(a,b))}{M(a,b,\widehat{r}(a,b))} - \frac{s}{b-a}, \tag{1.53}$$

where the partial derivatives of $M(a,b,r)$ (1.43) are

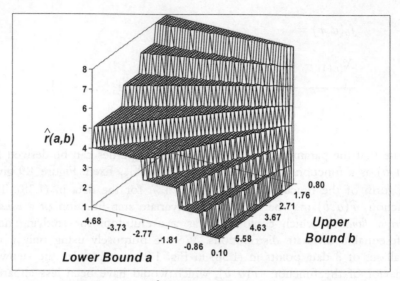

Fig. 1.9 The index function $\widehat{r}(a,b)$ given by Eq. (1.42) for the data in (1.38).

$$\frac{\partial}{\partial a}M(a,b,\widehat{r}(a,b)) = M(a,b,\widehat{r}(a,b)) \times \qquad (1.54)$$

$$\left\{ \sum_{j=1}^{\widehat{r}-1} \frac{Z_{(j)} - Z_{(\widehat{r})}}{(Z_{(\widehat{r})} - a)(Z_{(j)} - a)} \right\} < 0$$

and

$$\frac{\partial}{\partial b}M(a,b,\widehat{r}(a,b)) = M(a,b,\widehat{r}(a,b)) \times \qquad (1.55)$$

$$\left\{ \sum_{j=\widehat{r}+1}^{s} \frac{Z_{(j)} - Z_{(\widehat{r})}}{(b - Z_{(\widehat{r})})(b - Z_{(j)})} \right\} > 0.$$

A routine $BSearch$ has been developed utilizing (1.51), (1.53) and (1.55) to determine $\widetilde{b}\,(a)$ for fixed a, where

$$\widetilde{b}\,(a) = arg\max_{b > Z_{(s)}} \Big[G(a,b) \Big]. \qquad (1.56)$$

24

This routine follows a bisection approach (see, e.g., Press *et al.* (1989)) and is described in the next subsection. Having the routine $BSearch$ to determine $\widetilde{b}\ (a)$ for fixed a, we next compile a routine $ABSearch$ which determines $\widehat{a}$ and $\widetilde{b}\ (\widehat{a})$ such that

$$\widehat{a} = \underset{a < Z_{(1)}}{argmax} \left[G(a, \widetilde{b}\ (a)) \right].$$

The latter routine utilizes (1.52), (1.54) and is also based on a bisection approach. (It is described in the next subsection.) The routine $ABSearch$ evaluates the maximum of the likelihood, namely the RHS of (1.46), by successively utilizing $BSearch$ and yields the following ML estimators :

$$\widehat{a},\ \widehat{b} = \widetilde{b}\ (\widehat{a}),\ \widehat{m}(\widehat{a},\ \widehat{b}) = Z_{(\widehat{r}(\widehat{a},\ \widehat{b}))} \text{ and}$$

$$\widehat{n}(\widehat{a},\widehat{b}) = -\frac{s}{Log\{M(\widehat{a},\widehat{b},\widehat{r}(\widehat{a},\widehat{b})\}}$$

where $\widetilde{b}\ (\ \cdot\ )$ and $\widehat{r}(a,b)$ are defined in (1.56) and (1.42), respectively. For ease of implementation, the ML procedure above is summarized in Pseudo Pascal in the next subsection. We emphasize that the procedure — although straightforward — requires utilization of a number of variables and careful analysis of the consecutive steps and their interconnection.

1.5.2 ML estimation procedure in pseudo Pascal

The numerical routines below in Pseudo Pascal require separate algorithms to evaluate:

$M(a_k, b_k, r_k)$: Eq. (1.43), $G(a_k, b_k, r_k)$: Eq. (1.48),

$\dfrac{\partial}{\partial a} G(a_k, b_k, r_k)$: Eq. (1.52), $\dfrac{\partial}{\partial b} G(a_k, b_k, r_k)$: Eq. (1.53),

$\dfrac{\partial}{\partial a} M(a_k, b_k, r_k)$: Eq. (1.54) and $\dfrac{\partial}{\partial b} M(a_k, b_k, r_k)$: Eq. (1.55).

Output parameters of routines below are indicated in bold.

1.5.2.1 The search routine Bsearch

Let $G(a, b)$ be the function defined by (1.48). For a given value of the parameter a the set of discontinuities in the parameter b of the function $G(a, b)$ is a (finite) null-set and one could thus utilize the partial derivatives with respect to b (1.53) and (1.55) to determine an ascending search direction with respect to $G(a, b)$ for b. Define

$$B(a) = \underset{r \in \{1, \ldots, s-1\}}{Max} \left[f_b(a, r) \right], \tag{1.57}$$

where $f_b(a, r)$ are the discontinuity points given by (1.51). From the properties of $\widehat{r}(a, b)$ (1.42) mentioned at the beginning of Sec. 1.5.1, it follows that for $b > B(a)$ (*outside* the discontinuity locations) :

$$G(a, b) = Log \left\{ \prod_{i=2}^{s} \frac{b - Z_{(i)}}{b - Z_{(1)}} \right\} - sLog\{b - a\}$$

and

$$\frac{\partial}{\partial a} G(a, b) = \frac{s}{b - a} > 0. \tag{1.58}$$

Compare with the derivative (1.52). Hence, it follows from (1.58) that necessary conditions for a local maximum of $G(a, b)$ (i.e. $\frac{\partial}{\partial a}G(a, b) = 0$ and $\frac{\partial}{\partial b}G(a, b) = 0$) cannot be satisfied for $b > B(a)$. Thus, *BSearch* maximizing $G(a, b)$ as a function of b with a fixed, can be confined to the interval $(Z_{(s)}, B(a))$ only. The routine *BSearch* below evaluates $\widetilde{b}(a)$ (1.56), follows a bisection approach (see, e.g., Press *et al.* (1989)) and requires a separate algorithm to evaluate $B(a)$ (1.57).

$BSearch(a_k, \underline{Z}, \boldsymbol{b_k}, \boldsymbol{M_k}, \boldsymbol{r_k})$
$Step\,1:$ $l_k^b = Z_{(s)}$
$Step\,2:$ $u_k^b = B(a_k), b_k = \frac{l_k^b + u_k^b}{2},$
               $M_k = M(a_k, b_k, r_k), G_k = \frac{\partial}{\partial b}G(a_k, b_k, M_k, r_k)$
$Step\,3:$ $If\,Abs(G_k) \geq \delta\ then$
                  $If\,G_k < 0\ then\ u_k^b :\ = b_k\ Else\ l_k^b :\ = b_k$
       $Else\,Stop$

$Step\,4:$ $If\,(u_k^b - l_k^b) \geq \delta\,Goto\,Step\,2\;Else\,Stop$

1.5.2.2 The search routine ABSearch

Let as above $G(a, b)$ be the function defined by (1.48). For a given value of the parameter b the set of discontinuities in the parameter a of the function $G(a, b)$ is a null-set and one could thus utilize the partial derivatives with respect to a (Eqs. (1.52) and (1.54)) to determine an ascending search direction with respect to $G(a, b)$ for a. The routine $ABSearch$ starts by establishing an interval $[A, X_{(1)}]$ such that

$$\frac{\partial}{\partial a}G(A, \widehat{b}(A)) > 0, \qquad (1.59)$$

where $\widehat{b}(A)$ maximizes $G(A, b)$ as a function of b (and is calculated using the $BSearch$ routine in Sec. 1.5.1.1). To determine A in (1.59) one may utilize (1.52) and (1.54). From

$$\lim_{a \to -\infty}\left[\widehat{r}(a, b)\right] = s,$$

it follows that for any given b, there is a sufficiently small a such that

$$G(a, b) = Log\left\{\prod_{i=1}^{s-1}\frac{Z_{(i)} - a}{Z_{(s)} - a}\right\} - sLog\{b - a\}$$

and

$$\frac{\partial}{\partial a}G(a, b) = \sum_{i=1}^{s-1}\frac{Z_{(i)} - Z_{(s)}}{(Z_{(i)} - u)(Z_{(s)} - a)} + \frac{s}{b - a}. \qquad (1.60)$$

It thus follows from (1.60) that for any given b there exists an a sufficiently small such that $\frac{\partial}{\partial a}G(a, b) > 0$. So far we can only conjecture that an A satisfying (1.59) does exist. Numerical analyses support this conjecture. Having established the search interval $[A, X_{(1)}]$, the routine $ABSearch$ follows (analogously to the routine $BSearch$) a bisection approach (see, e.g., Press *et al.* (1989)) and evaluates the RHS of (1.46) by successively utilizing the routine $BSearch$.

$ABSearch(\underline{Z}, \boldsymbol{a_k}, \boldsymbol{b_k}, \boldsymbol{m_k})$

$Step\,1$: $u_k^a = Z_{(1)}, \, l_k^a = Z_{(1)} - (Z_{(s)} - Z_{(1)})$

$Step\,2$: $BSearch(l_k^a, \underline{Z}, \, \boldsymbol{b_k}, \boldsymbol{M_k}, \boldsymbol{r_k}),$

$\qquad\qquad$ $G_k = \frac{\partial}{\partial a}G(l_k^a, \, b_k, M_k, r_k)$

$Step\,3$: $If\,G_k < 0\,then$

$\qquad\qquad\qquad$ $u_k^a = l_k^a, l_k^a = l_k^a - (Z_{(s)} - Z_{(1)}), Goto\,Step\,2.$

$Step\,4$: $a_k = \frac{l_k^a + u_k^a}{2},$

$\qquad\qquad$ $BSearch(\,a_k, \underline{Z}, \, \boldsymbol{b_k}, \boldsymbol{M_k}, \boldsymbol{r_k}),$

$\qquad\qquad$ $G_k = \frac{\partial}{\partial a}G(\,a_k, b_k, M_k, r_k)$

$Step\,5$: $If\,Abs(G_k) \geq \delta\,then$

$\qquad\qquad\qquad$ $If\,G_k < 0\,then\,u_k^b := a_k\,Else\,l_k^b := a_k$

$\qquad\qquad$ $Else\,Goto\,Step\,7$

$Step\,6$: $If\,(u_k^a - l_k^a) \geq \delta\,then\,Goto\,Step\,4$

$\qquad\qquad$ $Else\,Goto\,Step\,7.$

$Step\,7$: $m_k = Z_{(r_k)}$

1.6 Solving for a and b using a Lower and Upper Quantile Estimate

We shall conclude our discussion of the triangular distribution by providing an appealing and smooth method of using quantile estimates to solve for a and b. Let Z be a triangular pdf with support $[a, b]$ and mode m with the pdf (1.6) and the cdf (1.11). As mentioned above, the recent popularity of the triangular distribution could perhaps be attributed to its use in uncertainty analysis packages such as @Risk (developed by the Palisade corporation). The package @Risk allows definition of a triangular distribution (via the function TRIGEN) by specifying a lower quantile a_p, a most likely value m and an upper quantile b_r such that

$$a < a_p \leq m \leq b_r < b. \qquad (1.61)$$

The latter avoids having to specify the lower and upper extremes a and b that by definition have a zero likelihood of occurrence (since, the triangular density equals zero at the bounds a and b). The software @Risk does not provide details, however, regarding how the bounds a and b are calculated given values for a_p, m and b_r. Keefer and Bodily (1983) formulated this problem in terms of two quadratic equations from which the unknowns a and b had to be solved numerically for the values $p = 0.05$ and $r = 0.95$. Although the numerical solution of their equations and their generalizations to other values of p and r do not pose any difficulties, we shall present here a slightly simplified version that only requires to solve numerically a single equation in the unknown quantity

$$q = \frac{m-a}{b-a}. \tag{1.62}$$

It follows from the cdf (1.11) that the quantity q equals the probability mass to the left of the mode m (and also equals the relative distance of the mode m to the lower bound a over the whole support $[a, b]$, which is unknown here).

From the definition of a_p, $(F(a_p|a, m, b) = p)$, we have from (1.11) and (1.62) that

$$a_p = a + (m-a)\sqrt{\frac{p}{q}}. \tag{1.63}$$

There is no direct relation between p and q here (contrary to the common notation when dealing with proportions and/or the binomial distribution), except that from (1.61) and (1.62) it follows that $0 < p < q < 1$. Solving for the parameter a from (1.63), yields using (1.62)

$$a \equiv a(q) = \frac{a_p - m\sqrt{\frac{p}{q}}}{1 - \sqrt{\frac{p}{q}}} < \frac{a_p - a_p\sqrt{\frac{p}{q}}}{1 - \sqrt{\frac{p}{q}}} = a_p. \tag{1.64}$$

(We use the notation $a(q)$ instead of a to emphasize that the lower bound a is a function of q, provided the p-th percentile a_p and the most likely value m are given.) Analogously to (1.64), we have for $m < b_r$ (using $b(q)$ in place of b):

$$b \equiv b(q) = \frac{b_r - m\sqrt{\frac{1-r}{1-q}}}{1 - \sqrt{\frac{1-r}{1-q}}} > \frac{b_r - b_r\sqrt{\frac{1-r}{1-q}}}{1 - \sqrt{\frac{1-r}{1-q}}} = b_r. \tag{1.65}$$

(Here we have from (1.61) and (1.62) that $1 - q > 1 - r > 0$).

Substituting $a(q)$ and $b(q)$ as given by (1.64) and (1.65) into (1.62), we arrive at the following *basic equation*

$$q = g(q) \tag{1.66}$$

where

$$g(q) = \frac{m - a(q)}{b(q) - a(q)} = \tag{1.67}$$

$$\frac{(m - a_p)\left(1 - \sqrt{\frac{1-r}{1-q}}\right)}{(b_r - m)\left(1 - \sqrt{\frac{p}{q}}\right) + (m - a_p)\left(1 - \sqrt{\frac{1-r}{1-q}}\right)}.$$

Observe the rather "structured" relation between $g(q)$ and q. Indeed, from its structure it immediately follows that $0 \le g(q) \le 1$ (as it should be since $g(q)$ represents the probability mass to the left of the mode m). In fact, setting $q = p$ ($q = r$) in the RHS of (1.67) yields $g(p) = 1$ ($g(r) = 0$). In addition, the denominator of the RHS (1.67) is "almost" a linear combination of the distances of the quantiles a_p and b_r from the mode m, with the weights that are determined by the quantile probability masses p and r and the probability mass q to the left of the mode m. In Chapter 4 (Sec. 4.3.3.3) we shall show that a *generalized* version of the Eq. (1.67) has a unique solution $q^* \in [p, r]$. One can solve numerically for q^* utilizing (1.66), the definition of $g(q)$ (1.67) and our favorite bisection method (see, e.g., Press *et al.* (1989)) with the starting interval $[p, r]$. After solving for the unique solution q^* of Eq. (1.66) one could calculate the associated lower and upper bounds $a(q^*)$ and $b(q^*)$ from Eqs. (1.64) and (1.65), respectively.

We shall illustrate the above procedure via the example:

$$a_p = 6.5, \, m = 7, \text{ and } b_r = 10.5, \, p = 0.10 \text{ and } r = 0.90. \tag{1.68}$$

Figure 1.10 depicts the function $g(q)$ (1.67) for the example above. Note that as stated above $g(p) = 1$, $g(r) = 0$ in this case and that the unique solution $q^* = 0.2198$ is the intersection of the function $g(q)$ with the positive diagonal of the unit square (indicated by a dotted line in Fig. 1.10). We calculated the value of q^* using the standard root finding algorithm GOALSEEK available in Microsoft Excel. Next, from (1.64) and (1.65) and utilizing $q^* = 0.2198$, we obtain for the lower and upper bounds

$$a(q^*) = 5.464 \text{ and } b(q^*) = 12.452,$$

respectively.

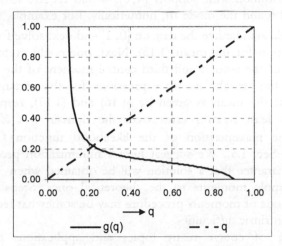

Fig. 1.10 The function $g(q)$ given by (1.67) with
$a_p = 6.5, m = 7$, and $b_r = 10.5, p = 0.10$ and $r = 0.90$.

1.7 Concluding Remarks

We have presented some details and properties of the triangular distribution which possibly have not been sufficiently addressed in the statistical literature. For example, to the best of our knowledge, the three-parameter ML method for the triangular distribution was first presented in Van Dorp and Kotz (2002b). The software BESTFIT developed by the Palisade corporation (which has been already available for a number of years now), however, does yield exactly the same estimates for the parameters a, m and b for the data in Table 1.2. Unfortunately, the authors do not provide specific details on their method for obtaining these estimates. On the other hand, another fitting software package called INPUT ANALYZER (developed by Rockwell Software) does *not* yield the same parameter estimates for the data in Table 1.2. (Again, no details are provided about the estimation procedure.)

    A careful reader would have noticed that the method of moments for the standard triangular distribution (1.4) with support $[0, 1]$ has not been explicitly discussed in this chapter due to its obvious simplicity. One may in fact directly solve for its the threshold parameter θ from the expression for the mean in (1.15). A three parameter method of moments procedure for a triangular distribution with support $[a, b]$, would require to solve for the bounds a and b and the mode m, numerically. For example, for a fixed a and b one could standardize the data on $[0, 1]$ and next solve for $\widehat{\theta}$ using the simple expression for the mean (1.15). Next, one could evaluate the least squares error of the second and third central moment of the standardized data utilizing the straightforward expressions for the variance and third moment about the mean as given by (1.16) and (1.17), respectively, and minimize this least squares error over the domain $a < Z_{(1)}$, $b > Z_{(s)}$ (similar to the maximization of the likelihood function $G(a, b)$ (1.48) introduced in Sec. 1.5). We suggest here a minimization procedure since there is no guarantee that a solution will be obtained when equating the first three sample moments to the theoretical ones. Steps used in the outlined methods of moments procedure may be somewhat tedious, but do not pose any intrinsic difficulties.

There are of course many topics and applications of triangular distributions which we were not able to cover in this chapter mainly due to space limitations. For completeness we are including in the bibliography citations of a number of papers not mentioned in the text that could be of interest to our diligent readers. These are appended by a star.

Chapter 2

Some Early Extensions of the Triangular Distribution

In this chapter we shall discuss in some detail three extensions of the triangular distribution that appeared prior to the 21-st century. The first one appeared in 1955 and is called the Topp and Leone distribution. It utilizes the left triangular distribution as its generating density in the same manner the Weibull distribution is generated from the exponential distribution. The second one is the trapezoidal distribution which perhaps can be considered the most natural extension of the three. To the best of our knowledge an early written record of the trapezoidal distribution in the modern statistical literature is from 1970. The third extension is a generalization of the symmetric triangular distribution. In Chapter 1 it was observed that the symmetric triangular distribution with support $[0, 1]$ arises from the average of two uniform random variables on $[0, 1]$. Hence, the third generalization we shall be considering herein involves the weighted average of n uniform random where the weights sum up to one, but are not necessarily the same. The corresponding cumulative distribution was derived as early as 1971. While for the Topp and Leone and trapezoidal distributions we shall provide some properties, only the cdf and pdf are obtained for the third one.

2.1 The Topp and Leone Distribution

In an early issue of the Journal of the American Statistical Association (JASA), in 1955, before the beginning of computer assisted statistical methodology, an interesting paper on a bounded continuous distribution by Topp and Leone (1955) has appeared which originally received little attention. The paper was resurrected by Nadarajah and Kotz (2003) and amplified by investigations of van Dorp and Kotz (2002a and b) dealing with the Two-Sided Power (TSP) distributions (to be discussed in Chapters 5, 6 and 8) and other alternatives to the beta distribution which, as we have already mentioned in Chapter 1, has been used in numerous applications since the beginnings of the 20-th century.

The construction of the Topp and Leone distribution is rather simple and is based on the observation that by raising an arbitrary cdf $F(x) \in [0,1]$ to an arbitrary power $\beta > 0$, a new cdf

$$G(x) = F^{\beta}(x) \tag{2.1}$$

emerges with one additional parameter. (This devise was used earlier by the Swedish engineer W. Weibull (1939) in the course of proposing his well-known Weibull distribution, that has achieved substantial popularity in the second part of the 20-th century, especially in reliability and biometrical applications.) Note that, from (2.1) we have for the pdf

$$g(x) = \frac{d}{dx} G(x) = \beta F^{\beta-1}(x) f(x), \tag{2.2}$$

where $f(x)$ is the pdf corresponding to the cdf $F(x)$. In the above construction the cdf $F(x)$ may be referred to as the *generating cdf.* Figure 2.1 plots the behavior of the multiplicative constant $\beta F^{\beta-1}(x)$ as a function of $F(x)$. The curves in Fig. 2.1 are evidently those of a power distribution pdf $\beta x^{\beta-1}$, which have the uniform cdf $F(x) = x, x \in [0,1]$, as its generating cdf. Also note that it follows from (2.2) (and visually from Fig. 2.1) that for $\beta > 1$ ($\beta < 1$) and for larger values of x (i.e. those values of x having the value of $F(x)$ close to 1) the multiplicative factor $\beta F^{\beta-1}(x) > 1$ ($\beta F^{\beta-1}(x) < 1$). The reverse assertion is also true for smaller values of x. The latter immediately implies that the central moments around 0 associated with the pdf $g(x)$ are strictly larger (smaller) than those associated with the pdf $f(x)$ when $\beta > 1$ ($\beta < 1$), i.e.

$$\int x^k g(x) dx \equiv \int x^k \beta F^{\beta-1}(x) f(x) dx > \tag{2.3}$$

$$\int x^k f(x) dx \Leftrightarrow \beta > 1, \; k = 1, 2, \ldots$$

Property (2.3) relates to the useful *stochastic dominance* property

$$G(x) = F^\beta(x) < F(x) \Leftrightarrow \beta > 1$$

(see, e.g. Clemen and Reilly (2001)).

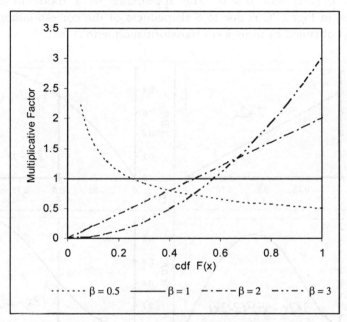

Fig. 2.1 The multiplicative factor $\beta F^{\beta-1}(x)$ in (9.0) as a function of the generating cdf $F(x)$.

Figure 2.2 presents a construction of the Topp and Leone distribution. The *generating cdf* of the Topp and Leone family is the cdf $(2x - x^2)$ with as its pdf being the left triangular density $2 - 2x, x \in [0, 1]$. They are displayed in Fig. 2.2B and Fig. 2.2A (as well as Fig. 1.2A in Chapter 1), respectively. Figures 2.2C and 2.2D plot the pdf and cdf of a one parameter Topp and Leone distribution constructed from the right triangular density $(2 - 2x)$, $x \in [0, 1]$, for $\beta = 3$. For the standard form of the Topp and Leone cdf one obtains from (2.2) and the cdf $(2x - x^2)$ of a left triangular distribution

$$F(x|\beta) = (2x - x^2)^\beta \tag{2.4}$$

35

with the pdf

$$f(x|\beta) = \beta(2 - 2x)(2x - x^2)^{\beta-1}, \tag{2.5}$$

where $x \in [0, 1]$ and $\beta > 0$. The appearance of a mode in the pdf presented in Fig. 2.2C is due to S-shapedness of the corresponding cdf in Fig. 2.2D obtained by using a cdf transformation with $\beta > 1$.

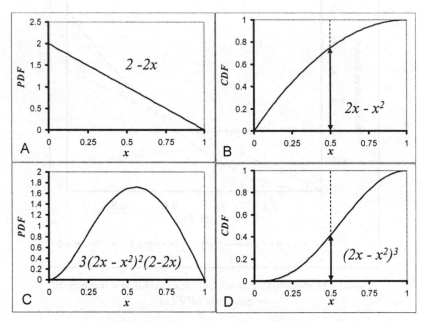

Fig. 2.2 Construction of Topp and Leone distribution from
a left triangular distribution. A: left triangular pdf; B: left triangular cdf;
C: Topp and Leone pdf with $\beta = 3$; D: Topp and Leone cdf with $\beta = 3$.

Figure 2.3 displays some examples of standard Topp and Leone distributions for different values of β. For $0 < \beta < 1$ ($\beta > 1$) the pdf (2.5) has a J-shaped (unimodal) form. For $\beta = 1$ (Fig. 2.3B), the Topp and Leone pdf (2.5) reduces to a left triangular pdf with support $[0, 1]$. The limiting distribution of (2.5) by letting $\beta \downarrow 0$ ($\beta \to \infty$) is a single point mass at 0 (at 1).

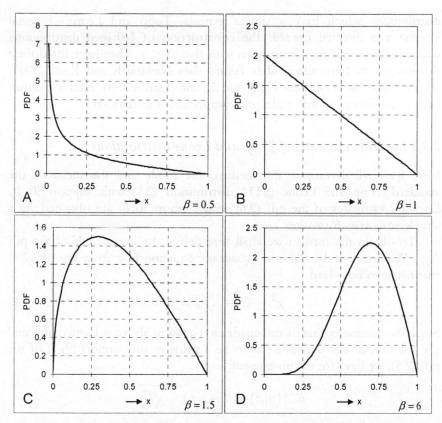

Fig. 2.3 Examples of Topp and Leone pdf's A: $\beta = 0.5$; B: $\beta = 1$; C: $\beta = 1.5$ D: $\beta = 6$.

Topp and Leone (1955) considered a non-standard version of (2.5) with support $[0, b]$ and the pdf

$$f(z|b, \beta) = \frac{2\beta}{b}(1 - \frac{z}{b})\left\{\frac{z}{b}\right\}^{\beta-1}\left\{2 - \frac{z}{b}\right\}^{\beta-1}, \qquad (2.6)$$

and cdf

$$F(z|b, \beta) = \left\{\frac{z}{b}\right\}^{\beta}\left\{2 - \frac{z}{b}\right\}^{\beta} \qquad (2.7)$$

where as above $\beta > 0$. Note that the expression for the pdf is more complicated than that of the cdf. Compare with the Weibull type distributions which have a similar genesis. Topp and Leone's original interest was directed towards the construction of J-shaped distributions utilizing similar cdf transformations with $0 < \beta < 1$. They have fitted their distribution to transmitter tubes failure data. Nadarajah and Kotz (2003) showed that the J-shaped Topp and Leone distributions exhibit bathtub failure rate functions with widespread applications in reliability.

2.1.1 Location estimates of Topp and Leone distributions

Here we shall investigate the median, the mode and the mean of the standard Topp and Leone (STL) distribution (2.5) with support $[0, 1]$. Location estimates of the pdf (2.6) with support $[0, b]$ are obtained via a simple scale transformation.

To obtain the median we shall first derive the inverse cdf of the pdf (2.5). We arrive at the following quadratic equation in x from the cdf (2.4) that needs to be solved

$$x^2 - 2x + \sqrt[\beta]{y} = 0. \tag{2.8}$$

Since the symmetry axis of the quadratic equation above is located at 1 and the solution of this equation has to be a value in the support $[0, 1]$ of the pdf (2.5), we have for the inverse cdf

$$F^{-1}(y|\beta) = 1 - \sqrt{1 - \sqrt[\beta]{y}},$$

and hence the median is

$$x_{0.50} = 1 - \sqrt{1 - \sqrt[\beta]{\frac{1}{2}}}. \tag{2.9}$$

For $\beta = 1$ the pdf (2.5) reduces to a left triangular pdf with a mode at 0. The mode of the pdf (2.5) for $\beta \neq 1$ follows by setting the derivative of (2.5) with respect to x, which is

$$2\beta(2x - x^2)^{\beta-2}\{2(\beta - 1)(1 - x)^2 - \beta(2x - x^2)\}, \tag{2.10}$$

to 0, yielding the following quadratic equation

$$(2\beta - 1)\, x^2 - 2(2\beta - 1)x + 2(\beta - 1) = 0 \qquad (2.11)$$

that needs to be solved. Note that from (2.10) it immediately follows the pdf (2.5) being strictly decreasing for $0 < \beta < 1$. Hence when solving the quadratic equation (2.11), we only need to consider the case $\beta \geq 1$. Similar to the quadratic equation (2.8) the symmetry axis here is also located at 1 and the only possible solution in the support $[0, 1]$ of (2.11) yields the mode

$$m = 1 - \frac{1}{\sqrt{2\beta - 1}}, \beta > 1. \qquad (2.12)$$

To derive an expression for the mean of a Topp and Leone pdf (2.5) we shall utilize here (for convenience of the derivation) the general form of the cumulative moments

$$M_k = \int_0^1 x^k \{1 - F(x)\} dx. \qquad (2.13)$$

For $k = 0$, one obtains $E[X] = M_0$. We then have utilizing the cdf (2.4):

$$M_k = \frac{1}{k + 1} - 2^{2\beta + k + 1} \int_0^{\frac{1}{2}} u^{\beta + k} \left\{1 - u\right\}^{\beta} du.$$

These cumulative moments M_k can be expressed in terms of the incomplete Beta function

$$B(x \mid a, b) = \frac{\Gamma(a + b)}{\Gamma(a)\Gamma(b)} \int_0^x p^{a-1}(1 - p)^{b-1} dp,$$

yielding

$$M_k = \frac{1}{k + 1} - 2^{2\beta + k + 1} \times \qquad (2.14)$$
$$\frac{\Gamma(\beta + k + 1)\Gamma(\beta + 1)}{\Gamma(2\beta + k + 2)} B(\frac{1}{2} \mid \beta + k + 1, \beta + 1).$$

Substituting $k = 0$ into (2.14) yields the expression for the mean

$$E[X] = M_0 = 1 - 4^{\beta} \frac{\Gamma(\beta+1)\Gamma(\beta+1)}{\Gamma(2\beta+2)}. \tag{2.15}$$

In particular for $\beta = \frac{1}{2}$ we have for a STL distribution

$$\mu_1' = \Gamma^2(1\frac{1}{2}) = \frac{\pi}{4}.$$

Figure 2.4 plots the mean (2.15), the median (2.9) and the mode (2.12) of the Topp and Leone distribution as a function of the parameter β.

Note that in Figure 2.4 the mean and median almost coincide for all values of $0 \leq \beta \leq 10$ and are both 0 for $\beta = 0$. The mode, however, remains 0 up to $\beta = 1$, but is larger than the median and mean in the vicinity of $\beta = 10$. Around the value of $\beta \approx 2.70$ the three location measures achieve practically the same value ≈ 0.52. Figure 2.4B provides a close-up of Fig. 2.4A for the range of $\beta \in [2.55, 2.80]$. From Figs. 2.4A and 2.4B we observe that for values of $\beta < \beta_1 \approx 2.624$ we have the ordering

$$\text{Mean} > \text{Median} > \text{Mode.} \tag{2.16}$$

Similarly, for values of $\beta > \beta_3 \approx 2.729$ we have the reverse ordering

$$\text{Mean} < \text{Median} < \text{Mode.} \tag{2.17}$$

Even though it follows from Fig. 2.4 that the orderings (2.16) and (2.17) prevail over the domain $\beta > 0$, it is worthwhile to note from Fig. 2.4B that for $\beta_1 < \beta < \beta_2 \approx 2.706$ (for $\beta_2 < \beta < \beta_3$) we have a different ranking

$$\text{Mode} < \text{Mean} < \text{Median} \, (\text{Mean} < \text{Mode} < \text{Median}) \tag{2.18}.$$

In fact, the rankings (2.16) and (2.17) are most common (see, e.g. Runnerburg (1978)) where the first one typically holds for right skewed distributions (a tail towards the right) and the second one for left skewed ones. It is perhaps less well known that the rankings (2.16) and (2.17) are only valid under certain regularity conditions (see, e.g., Basu and Dasgupta (1993)) and that other rankings (like the ones in (2.18)) are also possible.

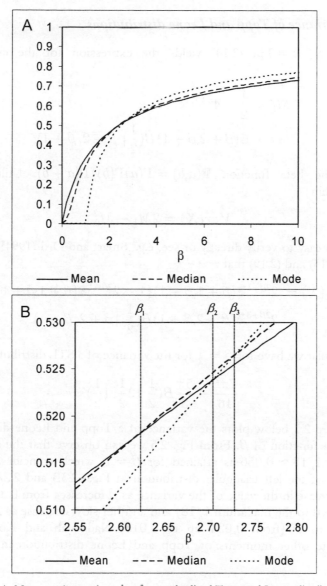

Fig. 2.4 A. Mean, median and mode of a standardized Topp and Leone distribution as a function of the parameter β. B. A close-up of Graph A for the range $\beta \in [2.55, 2.80]$, $\beta_1 \approx 2.624$, $\beta_2 \approx 2.706$ and $\beta_3 \approx 2.729$.

2.1.2 Variance of Topp and Leone distributions

Substituting $k = 1$ in (2.14) yields the expression for the cumulative moment

$$M_1 = \frac{1}{2} - 4^{\beta+1} \times \tag{2.19}$$

$$\mathbb{B}(\beta + 2, \beta + 1) B(\frac{1}{2} \mid \beta + 2, \beta + 1),$$

where the beta function $\mathbb{B}(a,b) = \Gamma(a)\Gamma(b)/\Gamma(a+b)$. Utilizing the relationship

$$Var(X) = 2M_1 - M_0^2$$

(which is easy to verify directly or see, e.g., Stuart and Ord (1994)) we have from (2.15) and (2.19) that

$$Var(X) = 2^{2\beta+1}\mathbb{B}(\beta+1, \beta+1)\{1 - 2^{2\beta-1}\mathbb{B}(\beta+1, \beta+1)\} - $$
$$2^{2\beta+3}\mathbb{B}(\beta+2, \beta+1)B(\frac{1}{2} \mid \beta+2, \beta+1).$$

In particular we have for $\beta = \frac{1}{2}$ for the variance of a STL distribution

$$\frac{7\pi^2}{16} - \frac{5\pi}{3}B(\frac{1}{2} \mid 2\frac{1}{2}, 1\frac{1}{2}).$$

Figure 2.5 below plots the variance of a Topp and Leone distribution (2.5) as a function of β. From Fig. 2.5 we can observe that the maximum variance $1/18 \approx 0.056$ is attained for $\beta = 1$, which coincides with the variance of the left triangular distribution in Figs. 2.3B and 2.2A. Note a rapid increase in the value of the variance as β increases from 0 to 1 (from the value 0 to the maximum $1/18$) and a rather slow decrease as β changes from 7 to 10 (from 0.024 up to 0.018). Nadarajah and Kotz (2003) investigate other moments of Topp and Leone distribution in a similar manner.

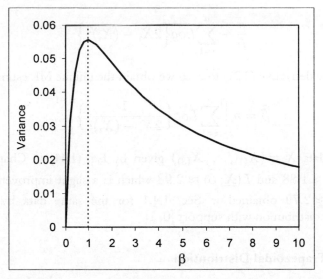

Fig. 2.5 Variance of a Topp and Leone distribution as a function of the parameter β.

2.1.3 Maximum likelihood estimation

We have seen at the beginning of this section that the structure of the pdf (2.5) and cdf (2.4) of the standard Topp and Leone distributions are appealingly direct. Moment expressions for Topp and Leone distributions, unfortunately, are rather cumbersome requiring numerical search routines when one would like to estimate its parameter β via the method of moments technique. Fortunately, the ML method does not present any intrinsic difficulties.

Let $\underline{X} = (X_1, \ldots, X_n)$ be an i.i.d. sample from a standard Topp and Leone distribution with the pdf (2.5) and support $[0, 1]$. By definition, the likelihood for $\underline{X}$ is

$$L(\underline{X}; \beta) = \beta^n \prod_{i=1}^{n} (2 - 2X_i)\left\{2X_i - (X_i)^2\right\}^{\beta-1} \qquad (2.20)$$

As is often the case, instead of maximizing the likelihood we shall equivalently maximize the log-likelihood. Taking the logarithm of (2.20) and calculating the derivative with respect to β we obtain

$$\frac{n}{\beta} + \sum_{i=1}^{n} Log\left\{ 2X_i - (X_i)^2 \right\} \tag{2.21}$$

Setting the derivative (2.21) to zero we obtain the unique ML estimator

$$\widehat{\beta} = n \left[\sum_{i=1}^{n} Log\left\{ \frac{1}{2X_i - (X_i)^2} \right\} \right]^{-1}.$$

For the data $X = (X_{(1)}, \ldots, X_{(8)})$ given by Eq. (1.41) in Chapter 1 we obtain $\widehat{\beta} \approx 1.88$ and $L(\underline{X}; \widehat{\beta}) \approx 2.98$ which is a slight improvement over $L(\underline{X}; \widehat{\theta}) \approx 2.79$ obtained in Sec. 1.4.1 for the same data by fitting a triangular distribution with support $[0, 1]$.

2.2 The Trapezoidal Distribution

Trapezoidal distributions have been advocated in risk analysis problems by Pouliquen as early as 1970 and more recently by Powell and Wilson (1997) and Garvey (2000). They have also found application as membership functions in the fuzzy set theory (see, e.g., Chen and Hwang (1992)). In that context they prominently occur in numerical geology and geostatistics (Bardosy and Fodor (2004)).

In a seminal paper in 1965, L.A. Zadeh observed that "more often than not, the classes of objects encountered in the real world do not have precisely defined criteria or membership". To specify a fuzzy set is to capture partial membership described by a membership function. A member ship functions corresponding to a fuzzy set is therefore a relaxation of the characteristic definition of a set (which has a rigid boundary). It is a gradual notion which is related to vagueness rather than uncertainty (see, Zadeh (1965)). The membership function of a fuzzy number is expressed by a continuous distribution on a bounded domain which is very often given by a triangular or trapezoidal distribution. A membership function of a fuzzy number is therefore similar in concept as the pdf of a random number. There exist rather *simple* rules for adding, subtracting and multiplying fuzzy numbers involving their membership functions which are quite different from convolutions or products, respectively of random variables (see, e.g., Bardosy and Fodor (2004) for details). It would seem that the fuzzy arithmetic is primarily developed for its mathematical convenience and simplicity, rather than adhering to the traditional views for propagating uncertainty. For example, the sum of two fuzzy numbers with triangular membership functions with parameters a_i, m_i, b_i, $i = 1, 2$, is again a fuzzy number with triangular membership function $a_1 + a_2, m_1 + m_2$, $b_1 + b_2$. Hence, fuzzy arithmetic avoids having to deal with the somewhat cumbersome calculation involved with the sum of two independent asymmetric triangular random variables (known as convolution), which does not follow a triangular density function. (See van Dorp and Kotz (2003b) for details.)

Our interest in trapezoidal distributions and their modifications stems mainly from the observations that many physical processes in nature, human body and mind (over time) reflect the form of the trapezoidal distribution. In this context, trapezoidal distributions have been used in medical applications, specifically the screening and detection of cancer (see, e.g., Kimmel and Gorlova (2003), Brown (1999) and Flehinger and Kimmel (1987)). Another domain for applications of the trapezoidal distribution is the applied physics arena (see, e.g. Davis and Sorenson (1969), Nakao and Iwaki (2000), Sentenac *et al.* (2000) and Straaijer and De Jager (2000)). Specifically, in the context of nuclear engineering, uniform and trapezoidal distribution have been assumed as models for observed axial distributions for burnup credit calculations (see Wagner and DeHart (2000) and Neuber (2000) for a comprehensive description).

The pdf of a trapezoidal distribution consists of three stages. (The first stage can be viewed as a growth-stage, the second corresponds to a relative stability and the third reflects a decline or decay). It seems appropriate to define the four-parameter trapezoidal pdf by

$$f_X(x|a, b, c, d) = C(a, b, c, d) \times \begin{cases} \frac{x-a}{b-a}, & \text{for } a \leq x < b \\ 1, & \text{for } b \leq x < c \\ \frac{d-x}{d-c}, & \text{for } c \leq x < d \end{cases} \quad (2.22)$$

where $a < b < c < d$ and the normalization constant is given by

$$C(a, b, c, d) = 2\{c + d - (a + b)\}^{-1}. \quad (2.23)$$

The corresponding cdf that follows from (2.22) is calculated to be:

$$F_X(x|a,b,c,d) = \begin{cases} \frac{b-a}{c+d-(a+b)} \left(\frac{x-a}{b-a}\right)^2, & \text{for } a \leq x < b \\ \frac{2x-(b+a)}{c+d-(a+b)}, & \text{for } b \leq x < c , \quad (2.24) \\ 1 - \frac{d-c}{c+d-(a+b)} \left(\frac{d-x}{d-c}\right)^2, & \text{for } c \leq x < d. \end{cases}$$

(see, also Garvey (2000), p. 103). The name "trapezoidal" reflects the shape of a graph of the pdf (See Fig. 2.6). Triangular and uniform distributions are special cases in the trapezoidal family.

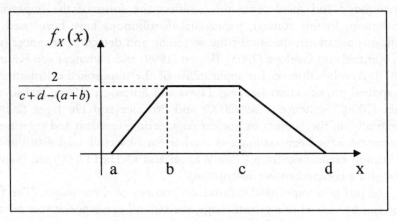

Fig. 2.6 Pdf of a Trapezoidal Distribution.

2.2.1 *Moments of the trapezoidal distribution*

Derivations of moments of the trapezoidal distribution (and some other properties as well) are straightforward and follow most naturally from the realization that the pdf (2.22) can be viewed as a mixture

$$f_X(x|\Theta) = \sum_{i=1}^{3} \pi_i f_{X_i}(x|\Theta), \quad \sum_{i=1}^{3} \pi_i = 1, \ \pi_i > 0, \quad (2.25)$$

of the three component densities:

$$\begin{cases} f_{X_1}(x|\Theta) = f_{X_1}(x|a,b) = \frac{2}{b-a}\frac{x-a}{b-a}, & \text{for } a \le x \le b \\ f_{X_2}(x|\Theta) = f_{X_2}(x|b,c) = \frac{1}{c-b}, & \text{for } b \le x \le c \quad (2.26) \\ f_{X_3}(x|\Theta) = f_{X_3}(x|c,d) = \frac{2}{d-c}\frac{d-x}{d-c} & \text{for } c \le x \le d \end{cases}$$

with the mixture probabilities

$$\begin{cases} \pi_1 = \frac{(b-a)}{(b-a)+2(c-b)+(d-c)}, \\ \pi_2 = \frac{2(c-b)}{(b-a)+2(c-b)+(d-c)}, \quad\quad (2.27) \\ \pi_3 = \frac{(d-c)}{(b-a)+2(c-b)+(d-c)}. \end{cases}$$

The densities $f_{X_1}(x|a,b)$ and $f_{X_3}(x|c,d)$ are a right (Fig. 1.2B) and a left triangular density (Fig. 1.2A), respectively. The density $f_{X_2}(x|b,c)$ is a uniform density with support $[b,c]$. Note that the mixing probabilities in (2.27) are proportional to the durations $(b-a)$, $(c-b)$ and $(d-c)$ of the three stages of the trapezoidal density (2.22), where the duration of the middle stage is multiplied by a factor of 2 compared to the other two stages.

Utilizing the component representation (2.25) of the pdf (2.22) and the mixture probabilities $\pi_i, i = 1,2,3$ (Eq. (2.27)), we arrive at

$$E[X^k|\Theta] = \pi_1 E[X_1^k|a,b] + \pi_2 E[X_2^k|b,c] + \pi_3 E[X_3^k|c,d]. \quad (2.28)$$

From the pdf's in (2.26) we obtain the moments around zero of the component variables:

$$E[X_1^k|a,b] = \sum_{i=0}^{k} \binom{k}{i} a^{k-i}(b-a)^i \frac{2}{2+i}, \quad\quad (2.29)$$

$$E[X_2^k|b,c,\alpha] = \frac{1}{(c-b)}\frac{c^{k+1}-b^{k+1}}{k+1}$$

$$E[X_3^k|c,d,n_3] = \sum_{i=0}^{k} \binom{k}{i}(c-d)^i d^{k-i}\frac{2}{2+i}.$$

Note, the symmetric analogy of $E[X_1^k|a,b]$ and $E[X_3^k|c,d]$. Numerical calculations of k-th moment $E[X^k|\Theta]$ given by (2.28) are quite innocuous employing the current advances in computer technology and utilizing the closed form expressions for the k-th moment of the component random variables X_1, X_2, X_3 (Eq. 2.29) as well as the mixture probabilities $\pi_i, i = 1, 2, 3$ (2.27). Deriving a closed form for the expression of the moments $E[X^k|\Theta]$ for $X \sim f_X(x|\Theta)$ (Eq. (2.28)) in its general form, although somewhat tedious, is quite straightforward and does not present intrinsic difficulties.

We shall derive the close form expressions for the first and second moment about zero. Substituting $k = 1$ in (2.29) we have for the component variables:

$$E[X_1|a, b] = \frac{a+b}{3} , E[X_2|b, c] = \frac{c+b}{2}, E[X_3|c, d] = \frac{c+d}{3}. \quad (2.30)$$

Next, the substitution of (2.30) and the mixture probabilities (2.27) into the general expression for the k-th moment (2.28) yields the following elegant formula for the mean of X

$$E[X|a, b, c, d] = \frac{1}{3} \frac{(c+d)^2 - cd - (a+b)^2 + ab}{c+d - (a+b)}. \quad (2.31)$$

Similarly, substituting $k = 2$ in (2.29) yields the second moments around zero of the component variables

$$E[X_1^2|a, b] = a^2 + \frac{2}{3}a(b-a) + \frac{1}{2}(b-a)^2 ,$$

$$E[X_2^2|b, c] = \frac{2}{c-b} \frac{c^3 - b^3}{3},$$

$$E[X_3^2|c, d] = d^2 - \frac{2}{3}d(d-c) + \frac{1}{2}(d-c)^2,$$

and the expression for second moment around zero of a trapezoidal distribution

$$E[X^2|a, b, c, d] = \frac{1}{6} \frac{(c^2 + d^2)(c+d) - (a^2 + b^2)(a+b)}{c+d - (a+b)}. \quad (2.32)$$

(see also Garvey 2000, p. 104). For $c = b$ ($a = b$ and $c = d$) in (2.31) and (2.32) we arrive at the first and second moments of a triangular distribution with support $[a, d]$ and a mode at b (a uniform distribution with support $[b, c]$).

2.2.2 Inverse cumulative distribution function

We shall derive the inverse of the cdf (2.24) to allow for direct sampling from the trapezoidal distribution using the inverse cumulative distribution technique. From (2.24) and again utilizing the component representation (2.25) of the pdf (2.22) we have

$$
F_X^{-1}(y|a, b, c, d) = \tag{2.33}
$$
$$
\begin{cases}
a + \sqrt{y(b-a)(c+d-a-b)}, & \text{for } 0 \leq y < \pi_1 \\
(1-y)\frac{a+b}{2} + y\frac{c+d}{2}, & \text{for } \pi_1 \leq y < 1 - \pi_3 \\
d - \sqrt{(1-y)(d-c)(c+d-a-b)}, & \text{for } 1 - \pi_3 \leq y < 1,
\end{cases}
$$

where the mixture probabilities π_i, $i = 1, \ldots, 3$ are given by Eq. (2.27). Note that, the second branch of the inverse cdf is a convex combination of the midpoints of the first and third stage of the trapezoidal distribution. As before, for $c = b$ ($a = b$ and $c = d$) in (2.33) and (2.27) we arrive at the inverse cdf of a triangular distribution with support $[a, d]$ and a mode at b (a uniform distribution with support $[b, c]$).

In Chapter 6, we shall generalize the trapezoidal distribution utilizing extensively the component representation (2.25).

2.3 A Linear Combination of Uniform Variables

Schmidt (1934) was possibly the first to notice explicitly that the symmetric triangular pdf (1.2) with support $[0, 1]$ follows as the distribution of the arithmetic average of two uniform random variables U_1 and U_2 on $[0, 1]$ given by (1.3) in Chapter 1. Hence, it seems quite natural to view a weighted average of uniform random variables, where the weights $\underline{w} = (w_1, \ldots, w_n)$ are positive but not necessarily equal, to be an extension of the symmetric triangular distribution with support $[0, 1]$. First, let the random variable X be defined as:

$$
X = \sum_{i=1}^{n} w_i U_i, \quad w_i > 0, \quad \sum_{i=1}^{n} w_i = 1, \tag{2.34}
$$

where U_i, $i = 1, \ldots, n$ are mutually independent uniform random variables with support $[0, 1]$. From the definition of X and the expressions

$$E[U_i] = \frac{1}{2} \text{ and } Var[U_i] = 1/12, \ i = 1, \ldots, n.$$

it follows immediately that

$$E[X] = \frac{1}{2} \text{ and } Var[X] = \frac{1}{12}\sum_{i=1}^{n} w_i^2. \tag{2.35}$$

The cdf of the variable X given by (2.34) was derived by Mitra (1971) and a couple of years later by Barrow and Smith (1979). It is given by a rather formidable sum:

$$F(x|n, \underline{w}) = \left(n!\prod_{i=1}^{n} w_i\right)^{-1} \sum_{v_1=0}^{1}\ldots\sum_{v_n=0}^{1}(-1)^{\sum_{i=1}^{n} v_i} \times \tag{2.36}$$

$$\left(x - \sum_{i=1}^{n} w_i v_i\right)^n 1_{[0,\infty)}\left(x - \sum_{i=1}^{n} w_i v_i\right)$$

with the pdf

$$f(x|n, \underline{w}) = \left\{(n-1)!\prod_{i=1}^{n} w_i\right\}^{-1} \sum_{v_1=0}^{1}\ldots\sum_{v_n=0}^{1}(-1)^{\sum_{i=1}^{n} v_i} \times \tag{2.37}$$

$$\left(x - \sum_{i=1}^{n} w_i v_i\right)^{n-1} 1_{[0,\infty)}\left(x - \sum_{i=1}^{n} w_i v_i\right)$$

where $1_{[0,\infty)}(z)$ is the indicator function :

$$1_{[0,\infty)}(x) = \begin{cases} 1 & z \in [0,\infty) \\ 0 & \text{elsewhere.} \end{cases}$$

Note that the variables v_i, $i = 1, \ldots, n$ in (2.36) and (2.37) are binary. Unfortunately, the proofs of Mitra (1971) and Barrow and Smith (1979) are — although quite ingenuous — geared towards mathematically oriented readers and are not easy to follow. Recently, Van Dorp (2004) utilized the cdf (2.36) for modeling statistical dependence caused by 'common risk factors' in uncertainty analysis applications. In these applications, the uniform random variables U_i are considered to be latent variables where the weights w_i indicates the amount contributed by a common risk factor i to the *aggregate risk* X. Latent variable models have found wide applications in scientific investigation particular in the behavioral sciences in the second part of the 20-th century (see, e.g., Bartholomew (1987)). Van Dorp (2004) devised an elementary proof of the cdf (2.36) that will described in a subsection below.

For $n = 2$ and $w_1 < 1 - w_1$ (or $w_1 < \frac{1}{2}$) the pdf (2.37) reduces to

$$f(x|2, \underline{w}) = \begin{cases} \frac{x}{w_1(1-w_1)} & 0 \le x < w_1 \\ \frac{1}{1-w_1} & w_1 \le x < 1 - w_1 \\ \frac{x-(1-w_1)}{w_1(1-w_1)} & 1 - w_1 \le x \le 1. \end{cases} \tag{2.38}$$

Figure 2.7 depicts the pdf of $X = w_1 U_1 + w_2 U_2$, where $w_1 = \frac{1}{3}$ and $w_2 = \frac{2}{3}$. From Fig. 2.7 and the structure of (2.38) it follows that the pdf in Fig. 2.7 is a symmetric trapezoidal distribution with $a = 0$, $b = \frac{1}{3}$, $c = \frac{2}{3}$ and $d = 1$ (see, Eq. (2.22)). Hence, similarly to the genesis of Schmidt (1934), determining that the average of two independent uniform random variables yields a symmetric triangular distribution on $[0, 1]$, we conclude that a weighted average of two independent uniform random variables (where the weights do not have to be equal) results in a symmetric trapezoidal distribution on $[0, 1]$ (this case to the best of our knowledge had not been well known).

By restricting the weights w_i to be equal to $1/n$ one evidently obtains a special case of the pdf (2.37). Introducing the notation

$$j = \sum_{i=1}^{n} v_i,$$

and noting that the number of different ways in which a sequence of n binary variables can sum up to j equals the binomial coefficient

$$\binom{n}{j} = \frac{n!}{j!\,(n-j)!}, \quad 0 \le j \le n,$$

the pdf (2.37) simplifies in this case to

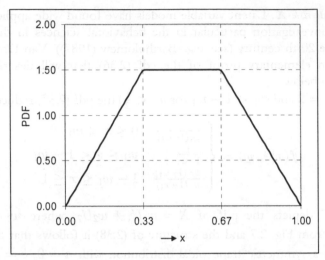

Fig. 2.7 Examples of the pdf's of (2.34) for $n = 2$ and $w_1 = \frac{1}{3}, w = \frac{2}{3}$.

$$f(x|n) = \frac{n^n}{(n-1)!} \sum_{j=0}^{\lfloor nx \rfloor} \binom{n}{j} (-1)^j (x - \frac{j}{n})^{n-1}, \qquad (2.39)$$

where $\lfloor nx \rfloor$ is the largest integer less than or equal to nx (also known as the entier function). The pdf (2.39) is often referred to as the Bates distribution (Bates (1955)) or the *rectangular mean* distribution. Bates (1955) derived her distribution for testing the hypothesis that a distribution is uniform $[0, 1]$ against the alternative that it represents a truncated exponential distribution on $[0, 1]$. Problems closely related to linear combinations of rectangular distributions (in connection with tabular differences) were investigated by Lowan and Laderman as early as (1939), whereas even earlier Irwin (1932) and Hall (1932) derived the distribution of the sum of n independent uniform random variables on $[0, 1]$. For $n = 2$, the pdf (2.39) reduces to the pdf (1.2) of a symmetric triangular pdf on $[0, 1]$ and hence for $n = 2m$ the pdf (2.39) represents the pdf of the sum of m independent *symmetric* triangular distributions with the support $[0, 1]$.

Figure 2.8 plots the pdf (2.37) of a random variable X given by (2.34) for $n = 4$ for different choices of the weights $w_i = 1, \ldots, 4$. Included in Fig. 2.8 is also the Bates distribution (with equal weights) which follows as

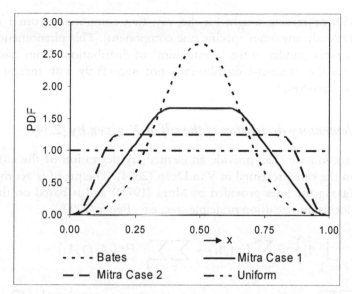

Fig. 2.8 Examples of the pdf's of (2.34) for $n = 4$. A: $w_i = \frac{1}{4}, i = 1, \ldots, 4$ (Bates);
B: $w_1 = 0.6, w_2 = 0.3, w_3 = 0.075, w_2 = 0.025$ (MB Case 1); C: $w_1 = 0.8, w_2 = 0.1$,
$w_3 = 0.075, w_2 = 0.025$ (MB Case 2); D: $w_1 = 1, w_i = 0, i = 2, 3, 4$ (Uniform).

the average of 4 independent uniform random variables or equivalently 2 independent symmetric triangular random variables with support $[0, 1]$. Note that the Bates distribution in Fig. 2.8 for $n = 4$ already takes on a bell-shaped form. A *standardized* Bates distribution with $n = 12$ and zero mean and variance 1 is commonly used for generation of standard normal variables in numerous packages. Note that the distributions indicated by 'Mitra Case 1' and 'Mitra Case 2' have a flat 'top' similar to the trapezoidal distribution in Fig. 2.6 and are also symmetric (in accordance with $E[X] = \frac{1}{2}$ in Eq. (2.35)). The pdf 'Mitra Case 2' in Fig. 2.8 assigns a higher weight to U_1 than the pdf 'Mitra Case 1' resulting in a larger variance. The maximum variance of X is attained for the limiting case where one of the weights w_i gets the maximal weight 1 and the other ones are all zero. In that case the pdf of X (2.34) reduces to a uniform pdf also depicted in Fig. 2.8. The minimum variance case of X is the Bates pdf with *equal* weights. We thus have a 'continuum' of continuous distributions in a finite domain with monotonically decreasing variances starting from the uniform distribution to the Bates distribution (2.39) via distributions of the form

(2.37) with decreasing weight for the, say, first component from 1 to $1/n$ (or equivalently any other specific one component). This phenomenon is in a certain sense similar to the "continuum" of distributions from Cauchy to normal via the Student-t distributions not necessarily with integer values degrees of freedom.

2.3.1 Elementary derivation of the cdf of X given by (2.34)

In this section we shall provide an elementary derivation of the cdf (2.34) based on the one presented in Van Dorp (2004). The proof is geometric in nature (along the lines provided by Mitra (1971)) and is based on the well-known inclusion-exclusion principle (see, e.g., Feller (1990))

$$Pr\left\{ \bigcup_{i=1}^{n} A_i \right\} = \sum_{i=1}^{n} Pr(A_i) - \sum\sum_{i<j} Pr\{A_i \cap A_j\} + \qquad (2.40)$$

$$\sum\sum\sum_{i<j<k} Pr\{A_i \cap A_j \cap A_k\} - \ldots + (-1)^{n-1} Pr\left\{ \bigcap_{i=1}^{n} A_i \right\},$$

for arbitrary not necessarily disjoint events $A_1, \ldots, A_n$, in the standard notation involving the union operator $\cup$ and the intersection operator $\cap$ of events. Mitra (1971) did not link his proof to the inclusion-exclusion principle above. Barrow and Smith's (1979) proof involved concepts related to the theory of splines which may be less well known to a probabilist. (This subsection could be skipped in the first reading of the book without losing continuity of exposition).

Let $C^n = \{u \mid 0 \le u_i \le 1\}$ be the unit hyper-cube in $\mathbb{R}^n$. Let $v = (v_1, \ldots, v_n)$, $v_i \in \{0, 1\}$, be a vertex (or corner point) of the unit hyper-cube C^n and define the simplex $S_{\underline{v}}(x)$ at the vertex $\underline{v}$ as:

$$S_{\underline{v}}(x) = \left\{ \underline{u} \mid \sum_{i=1}^{m} w_i u_i \le x, u_i \ge v_i, i = 1, \ldots, m \right\}, \qquad (2.41)$$

where $x \ge 0$, $w_i > 0$, $\sum_{i=1}^{m} w_i = 1$. Let $\underline{0} = (0, \ldots, 0)$ be the origin vertex of the unit hyper-cube C^n and let $\underline{e}^i = (e_1, \ldots, e_n)$, $i = 1, \ldots, n$, be its orthogonal unit vectors, i.e.

$$e_i = 1, e_j = 0, \; j = 1, \ldots, n, j \neq i. \tag{2.42}$$

For $n = 2$, C^2 reduces to the unit square and a simplex becomes a right angular triangle. For example, Fig. 2.9A displays the unit square C^2 and the simplex $S_{\underline{0}}(x_1)$ (Eq. (2.41)) with the corner points

$$\{(0,0), (\frac{x_1}{w_1}, 0), (0, \frac{x_1}{w_2})\}$$

for a particular value $x_1, 0 \le x_1 \le 1$. Figure 2.9B, displays $S_{\underline{0}}(x_2)$ for a value $x_2 > x_1$ and also depicts the simplex (or triangle) $S_{\underline{e}^1}(x_2)$ with corner points

$$\{(1,0), (\frac{x_2}{w_1}, 0), (1, \frac{x_2 - w_1}{w_2})\}$$

(*outside the unit square*) and the triangle $S_{\underline{e}^2}(x_2)$

$$\{(0,1), (\frac{x_2 - w_2}{w_1}, 1), (0, \frac{x_2}{w_2})\},$$

(*also outside the unit square*), respectively. To evaluate the cdf of X given by (2.34) for the value x_1 in Fig. 2.9A one simply needs to evaluate the area of the triangle $S_{\underline{0}}(x_1)$. Similarly, to evaluate the cdf of X for the value x_2 in Fig. 2.9B one calculates the area of the triangle $S_{\underline{0}}(x_1)$, *but subtracts the areas of the triangles with their right angles* at the unit vertices $\underline{e}^1$ and $\underline{e}^2$, i.e. $S_{\underline{e}^1}(x_2)$ and $S_{\underline{e}^2}(x_2)$.

Figure 2.10 presents similar situations for the case $n = 3$. Figure 2.10A displays the unit cube C^3 and the simplex $S_{\underline{0}}(x_1)$ (Eq. (2.41)). Note that in Fig. 2.10A, for this particular value $0 < x_1 < 1$ only the simplex $S_{\underline{0}}(x_1)$ at the origin $\underline{0} = (0,0,0)$ is a non-empty set since it is the only corner point of a total of 8 points of the cube C^3 that is an element of the half space

$$\{\underline{u} \mid \sum_{i=1}^{m} w_i u_i \le x_1\}. \tag{2.43}$$

When the value of x increases in (2.41), additional corner points $\underline{v}$ of the unit cube will join the half-space (2.43) resulting in additional non-empty simplecies at those points.

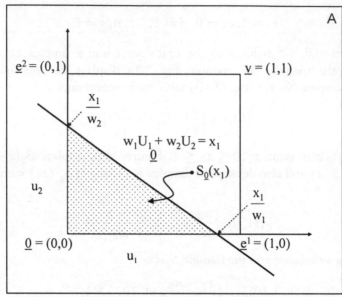

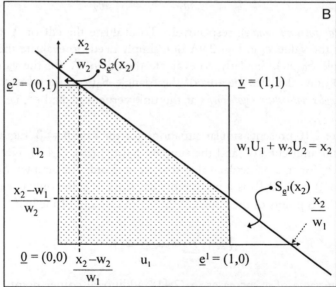

Fig. 2.9 A: Evaluating $F(x_1) = Pr(X \leq x_1)$ (Eq. (2.36)) for $n = 2$;
B: Evaluating $F(x_2) = Pr(X \leq x_2)$ (Eq. (2.36)) for $n = 2$.

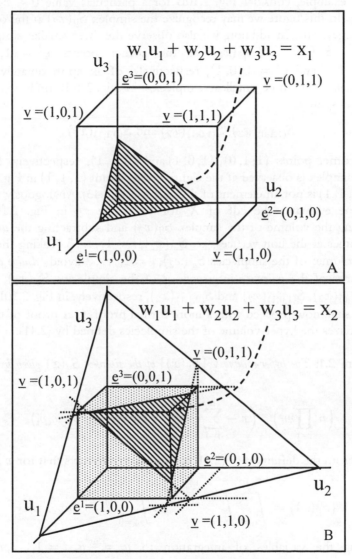

Fig. 2.10 A: Evaluating $F(x_1) = Pr(X \leq x_1)$ (Eq. (2.36)) for $n = 3$;
B: Evaluating $F(x_2) = Pr(X \leq x_2)$ (Eq. (2.36)) for $n = 3$.

For example, consider Fig. 2.10B for a particular value $0 < x_2 < 1$, $x_2 > x_1$. In this figure we may recognize the simplex $S_{\underline{0}}(x_2)$ at the origin $\underline{0}$ as the largest one. In addition, we also observe the three smaller simplecies $S_{\underline{e}^1}(x_2)$, $S_{\underline{e}^2}(x_2)$ and $S_{\underline{e}^3}(x_2)$ at the corner points $\underline{e}^1 = (1,0,0)$, $\underline{e}^2 = (0,1,0)$ and $\underline{e}^3 = (0,0,1)$, respectively, of an approximately equal size. Finally, the three smallest simplecies in Fig. 2.10B (indicated with dotted lines) are :

$$S_{(1,1,0)}(x_2), S_{(1,0,1)}(x_2) \text{ and } S_{(0,1,1)}(x_2)$$

at the corner points $(1,1,0)$, $(1,0,1)$ and $(0,1,1)$, respectively. Notice that no simplex is observed at the eighth corner-point $(1,1,1)$ in Fig. 2.10B since $(1,1,1)$ is not an element of the half space (2.43). Analogously to Fig. 2.9B, one evaluates the cdf of X for the value x_2 in Fig. 2.10B by calculating the volume of the simplex $S_{\underline{0}}(x_2)$ and subtracting the *union* of the volumes at the unit vertices, i.e. $S_{\underline{e}^i}(x_2), i = 1, \ldots, 3$. Taking the *union* of the volumes of the simplecies $S_{\underline{e}^i}(x_2), i = 1, \ldots, 3$ avoids *double counting* of volumes of the pairwise intersections of the simplecies $S_{\underline{e}^i}(x_2)$, which are $S_{(1,1,0)}(x_2)$, $S_{(1,0,1)}(x_2)$ and $S_{(0,1,1)}(x_2)$, respectively, in Fig. 2.10B.

We shall now proceed with a more formal proof. Our proof of the cdf (2.36) utilizes the hyper-volume of the simplecies defined by (2.41).

Theorem 2.1: *The hyper-volume $V\{S_{\underline{v}}(x)\}$ of the simplex $S_{\underline{v}}(x)$ given by (2.41) equals*

$$\left(n! \prod_{i=1}^{n} w_i\right)^{-1} \left(x - \sum_{i=1}^{n} w_i v_i\right)^n \cdot 1_{[0,\infty)}\left(x - \sum_{i=1}^{n} w_i v_i\right). \qquad (2.44)$$

Proof: From the definition of (2.41) it immediately follows that for $x \geq 0$

$$V\{S_{\underline{0}}(x)\} = \int_{u_1=0}^{\frac{x}{w_1}} \int_{u_2=0}^{\frac{x}{w_2} - \sum_{i=1}^{1} \frac{w_i}{w_2} u_i} \cdots \int_{u_n=0}^{\frac{x}{w_n} - \sum_{i=1}^{n-1} \frac{w_i}{w_n} u_i} du_n \ldots du_1. \qquad (2.45)$$

Changing the variables of integration to $y_i = \frac{w_i}{x} u_i, i = 1, \ldots, n$, the integral in (2.45) is simplified to

$$V\{S_{\underline{0}}(x)\} = \frac{x^n}{n \prod_{i=1}^{n} w_i} \int_{y_1=0}^{1} \int_{y_2=0}^{1-\sum_{i=1}^{1} y_i} \cdots \int_{y_n=0}^{1-\sum_{i=1}^{n-1} y_i} dy_n \ldots dy_1. \quad (2.46)$$

For $n = 2$, we evidently have for the value of the integral in (2.46)

$$\int_{y_1=0}^{1} \int_{y_2=0}^{1-y_1} dy_2 dy_1 = \int_{y_1=0}^{1} (1 - y_1) dy_1 = \frac{1}{2}.$$

For $n = 3$, one obtains

$$\int_{y_1=0}^{1} \int_{y_2=0}^{1-y_1} \int_{y_3=0}^{1-y_1-y_2} dy_3 dy_2 dy_1 = \quad (2.47)$$

$$\int_{y_1=0}^{1} \int_{y_2=0}^{1-y_1} (1 - y_1 - y_2) dy_2 dy_1 =$$

$$\frac{1}{2} \int_{y_1=0}^{1} (1 - y_1)^2 dy_1 = \frac{1}{2}\frac{1}{3} = \frac{1}{6}.$$

Generalizing Eq. (2.47) to an arbitrary integer $n \geq 1$ it immediately follows from (2.46) and the fact that $S_{\underline{0}}(x) = \emptyset$ for $x < 0$ that

$$V\{S_{\underline{0}}(x)\} = \frac{x^n}{n! \prod_{i=1}^{n} w_i} \cdot 1_{[0,\infty)}(x). \quad (2.48)$$

Once more changing variables $z_i = u_i - v_i$, $i = 1, \ldots, m$, we arrive, utilizing (2.41), at

$$V\{S_{\underline{v}}(x)\} = V\{S_{\underline{0}}(x - \sum_{i=1}^{n} w_i v_i)\}. \quad (2.49)$$

The theorem now follows from (2.48) and (2.49) via simple algebraic manipulations. $\qquad\qquad\qquad\qquad\qquad\qquad\qquad\qquad\qquad\qquad\qquad\square$

Theorem 2.2: *The cdf of the weighted linear combination X given by (2.34), where $U_i, i = 1, \ldots, n$, are independent uniform random variables with support $[0, 1]$ and*

$$\sum_{i=1}^{n} w_i = 1, w_i > 0,$$

is given by $F(x|n, \underline{w})$ *in Eq.* (2.36).

Proof: The support of X follows from (2.34) to be $[0, 1]$. From (2.34), the independence of the random variables U_i, $i = 1, \ldots, n$ and by generalizing the earlier observations in connection with Figs. 2.9B and 2.10B to $\mathbb{R}^n$, we obtain directly

$$Pr(X \leq x) = V\{S_{\underline{0}}(x)\} - V\left\{\bigcup_{i=1}^{n} S_{\underline{e}^i}(x)\right\}. \tag{2.50}$$

Invoking the inclusion-exclusion principle (Eq. (2.40)) we arrive at:

$$V\left\{\bigcup_{i=1}^{n} S_{\underline{e}^i}(x)\right\} = \sum_{i=1}^{n} V\{S_{\underline{e}^i}(x)\} - \tag{2.51}$$

$$\sum\sum_{i<j} V\{S_{\underline{e}^i}(x) \cap S_{\underline{e}^j}(x)\} +$$

$$\sum\sum\sum_{i<j<k} V\{S_{\underline{e}^i}(x) \cap S_{\underline{e}^j}(x) \cap S_{\underline{e}^k}(x)\} - \ldots$$

$$+ (-1)^{n-1} V\left\{\bigcap_{i=1}^{n} S_{\underline{e}^i}(x)\right\}.$$

Utilizing the definition of the simplex $S_{\underline{v}}(x)$ as given by (2.41) it follows that the intersections of the simplecies $S_{\underline{e}^i}(x)$ in (2.51) are all of the following form

$$\bigcap_{i \in I} S_{\underline{e}^i}(x) = S_{\underline{v}}(x), \tag{2.52}$$

where $I \subset \{1, \ldots, m\}$ and $\underline{v} = \sum_{i \in I} \underline{e}^i$. (For example, $S_{(1,1,0)}(x_2)$ in Fig. 2.10B is the intersection of $S_{(1,0,0)}(x_2)$ and $S_{(0,1,0)}(x_2)$.) Finally, from (2.52), (2.51) and (2.50) we easily conclude that

$$Pr(X \leq x) = \sum_{v_1=0}^{1} \cdots \sum_{v_n=0}^{1} (-1)^{\sum_{i=1}^{n} v_i} V\{S_{\underline{v}}(x)\}.$$

The proof of Theorem 2.2 now follows from Theorem 2.1. $\square$

From the proof of Theorem 2.2 it follows that an efficient method to evaluate the distribution in (2.36) for a particular value of x and a given set of weights $\underline{w} = (w_1, \ldots, w_m)$ is to develop a recursive algorithm enumerating all vertices $\underline{v}$ of the hyper-cube C^n and then to evaluate the hyper-volume (Eq. (2.44)) of the simplex at each vertex $\underline{v}$ when a vertex is visited by the algorithm. Such an efficient algorithm for evaluation of the cdf (2.36) facilitates its application in Monte Carlo based uncertainty analyses. The algorithm is presented below in Pseudo Pascal.

2.3.2 Algorithm for evaluating the cdf given by (2.36)

The procedure $CalcCDF(F, x, m, \underline{w})$ below evaluates the value F of the cdf (2.36) of the random variable X given by (2.34) by making a call to the recursive procedure

$$VisitVertices(F, x, i, \underline{v}, n, \underline{w}, \Pi).$$

As above the algorithm uses functions

$$ProductWeights(\underline{w}, n)$$

to calculate $\Pi = \prod_{i=1}^{n} w_i$

$$SumElements(\underline{v}, n)$$

to calculate $\Sigma = \sum_{i=1}^{n} v_i$, and

$$SumProducts(\underline{v}, \underline{w}, n)$$

to calculate $\psi = \sum_{i=1}^{n} w_i v_i$ (Eq. (2.36)).

The algorithm consists of two procedures each containing three and four steps, respectively:

$VisitVertices(F, x, i, \underline{v}, m, \underline{w}, \Pi);$
$Step\ 1:$ $if\ (i < m)\ then$
                 $v_i: = 0;\ VisitVertices(F, x, i, \underline{v}, n, \underline{w}, \Pi);$
                 $v_i: = 1;\ VisitVertices(F, x, i, \underline{v}, n, \underline{w}, \Pi);$
$Step\ 2:$ $\Sigma: = SumElements(\underline{v}, n);$
             $\psi: = SumProducts(\underline{v}, \underline{w}, n);$
$Step\ 3:$ $If\ (x - \psi) > 0\ then\ F: = F + (-1)^{\Sigma}\left\{\frac{x-\psi}{n!\Pi}\right\}$

$CalcCDF(G, x, m, w);$
$Step\ 1:$ $If\ x \leq 0\ then\ F: = 0;\ Stop;$
$Step\ 2:$ $If\ x \geq 1\ then\ F: = 1;\ Stop;$
$Step\ 3:$ $\Pi: = ProductWeights(\underline{w}, n);$
$Step\ 4:$ $VisitVertices(F, x, 1, \underline{v}, n, \underline{w}, \Pi);$

2.4 Concluding Remarks

This chapter covers selected topics constituting a transition from the classical contributions related to the triangular distributions which have originated in the 18-th century (strongly influenced by combinatorial problems) to the modern late 20-th century investigations in the area of bounded univariate continuous distributions motivated mainly by engineering and statistical considerations. Some statistical tools used in this chapter will be useful when studying more complex distributions described in Chapters 3-8. Careful attention given to this chapter thus may facilitate understanding of the material in the later chapters.

Chapter 3

The Standard Two-Sided Power Distribution

Similarly to the well known Gaussian or normal distribution, the beta distribution is represented by a smooth function. Whereas a "peaked" alternative for the normal distribution has been available for quite some time in the form of the Laplace distribution, a flexible "peaked" alternative for the beta distribution was lacking in the 20-th century. Smoothness of density curves may be an attractive mathematical property, but it does not necessary have to be dictated by the uncertainty of the phenomenon one is attempting to describe. In particular, financial data has been shown to exhibit "peaked" histograms. Ironically, mathematical idealization calls for smoothness while the real world often exhibits peakedness. The standard two-sided power (STSP) family of distributions introduced in this chapter can be considered a "peaked" alternative to the beta family. Properties of the distribution are investigated and a maximum likelihood estimation procedure for its two parameters is derived. The flexibility of the family in comparison with that of the beta family, is analyzed.

3.1 Introduction: The Leading Example

We shall introduce the STSP distribution using data for monthly interest rates for 30-year conventional mortgage interest rates over the period 1971-2003. The interest rate after month k is denoted by i_k; one of the simplest financial engineering models for the random behavior of the interest rates is the multiplicative model:

$$i_{k+q} = i_k \, \epsilon_{k,q}, \tag{3.1}$$

where $q = 1, 2, \dots$ and $\epsilon_{k,q}$ are i.i.d. random variables (see, e.g., Leunberger (1998)).

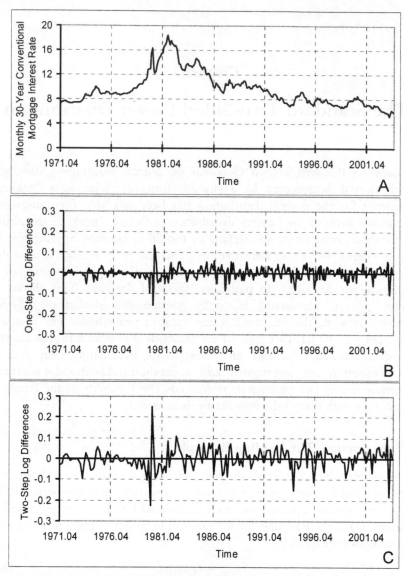

Fig. 3.1 Monthly 30-Year conventional mortgage interest rate data for the years 1971 - 2003;
A: Time series of interest rates; B: Time series of one-step log differences;
C: Time series of two-step log differences.

The time series of the monthly interest rates i_k consisting of 392 data points is displayed in Fig. 3.1A. From (3.1) we have

$$Ln(\epsilon_{k,1}) = Ln(i_{k+1}) - Ln(i_k). \tag{3.2}$$

Figure 3.1B depicts the time series of the one-step (or monthly) log differences $Ln(\epsilon_{k,1})$ consisting of 391 data points. Table 3.1 contains the values of the auto-correlation function (ACF)

$$ACF(\lambda, 1) = Corr[Ln(\epsilon_{k+\lambda,1}), Ln(\epsilon_{k,1})] \tag{3.3}$$

with lags $\lambda = 1, \ldots, 6$ together with the $LBQ(\lambda)$ statistic (see Ljung and Box (1978)) and their p-values for testing the null hypothesis that the auto-correlations for all lags up to lag λ are zero. The $LBQ(\lambda)$ statistic is chi-squared distributed with λ degrees of freedom (see, e.g., Tsay (2002)). Tsay (2002) suggests that the lag $\lambda = 6 \approx Ln(392) = 5.969$ performs better as far as statistical power is concerned than other values of λ. Hence, Table 3.1 contains the values of the $LBQ(\lambda)$ statistic up to and including $\lambda = 6$.

Table 3.1 Auto-correlation function, Ljung-Box Q statistic and p-values
for one-step log differences $Ln(\epsilon_{k,1})$ (Eq. (3.2)) and
two-step log differences $Ln(\epsilon_{k,2})$ (Eq. (3.3)) with Lags $\lambda = 1,\ldots,6$.

Lag	One-Step Log Differences			Two-Step Log Differences		
	ACF	LBQ	p-value	ACF	LBQ	p-value
1	0.414	67.646	2.0E-16	-0.020	0.083	0.773
2	-0.073	69.740	7.2E-16	0.038	0.371	0.831
3	-0.084	72.502	1.2E-15	-0.029	0.537	0.911
4	0.047	73.382	4.4E-15	0.028	0.697	0.952
5	0.028	73.687	1.7E-14	0.097	2.626	0.757
6	0.037	74.221	5.6E 14	0.005	2.632	0.853

From the p-values associated with the one-step log differences one immediately concludes that these null hypotheses (all lags up to lag λ being zero) are all rejected for lags $\lambda = 1, \ldots, 6$.

Figure 3.1C depicts the time series of the two-step (or bi-monthly) log-differences ($q = 2$ in Eq. (3.1))

$$Ln(\epsilon_{k,2}) = Ln(i_{k+2}) - Ln(i_k), \tag{3.4}$$

totaling 196 data points and the last 3 columns of Table 3.1 contain the values of the auto correlation function

$$ACF(\lambda, 2) = Corr[Ln(\epsilon_{k+\lambda,2}), Ln(\epsilon_{k,2})]$$

with the same lags $\lambda = 1, \ldots, 6$ together with the LBQ(λ) statistics and their p-values. Note that from these p-values it follows that in this case the null hypotheses (i.e. that auto-correlations for all the lags up to lag λ being zero) would be accepted for lags $\lambda = 1, \ldots, 6$. Hence, we may assume here that the random variables $\epsilon_{k,2}$ are indeed i.i.d.. This allows us the use of standard maximum likelihood (ML) procedures. (If a reader is not comfortable with this assumption additional modeling using the Auto-Regressive Conditional Heteroscedastic (ARCH) time series model devised by R.F. Engle (a 2003 Nobel Laureate in Economics) in 1982 may be used to construct an i.i.d. sequence from $\epsilon_{k,2}$ — see Sec. 6.5 in Chapter 6 for further details).

Figure 3.2 depicts the two-parameter Gaussian distribution together with an empirical pdf of the two-step log differences depicted in Fig. 3.1C. Similar to the analysis in Klein (1993) (who studied interest rate data on 30-year treasury bond data from 1977 to 1990), Fig. 3.2 shows that the empirical pdf of the financial data is by far too peaked to be modeled by a normal pdf. Figure 1A also displays the three-parameter asymmetric Laplace (AL) pdf

$$f(x|\mu, \kappa, \sigma) = \begin{cases} \frac{\sqrt{2}}{\sigma} \frac{\kappa}{1+\kappa^2} exp\left\{ -\sqrt{2} \frac{1}{\sigma\kappa}(\mu - x) \right\}, & \text{for } x < \mu \\ \frac{\sqrt{2}}{\sigma} \frac{\kappa}{1+\kappa^2} exp\left\{ -\sqrt{2} \frac{\kappa}{\sigma}(x - \mu) \right\}, & \text{for } x \geq \mu, \end{cases} \quad (3.5)$$

where $\mu \in \mathbb{R}$ and $\sigma, \kappa > 0,$ suggested by Kozubowski and Podgórski (1999) to capture such a peak (which appears to be a characteristic of financial data). Both the Gaussian and asymmetric Laplace distributions in Fig. 3.2 were fitted utilizing a ML procedure. For ML estimates of Gaussian parameters see, e.g., Mood *et al.* (1974). Kotz *et al.* (2002) provide a ML procedure for the asymmetric Laplace distribution.

Both the Gaussian and asymmetric Laplace distribution in Fig. 3.2 have unbounded support whereas the range of the two-step log difference in Fig. 3.1C is finite (and so is the range of the empirical pdf in Fig. 3.2) and equals $[-0.224, 0.250]$, not even covering a unit distance. We suggest to use continuous distributions with bounded support to model the uncertainty of the data in Fig. 3.1C. Figure 3.3 depicts the ML fitted pdf of a shifted standard beta distribution with unit support $[-0.5, 0.5]$ (and a sufficient safety margin for the range $[-0.224, 0.250]$ of the two-step log differences) and pdf

$$f(x|a, b) = \frac{\Gamma(a+b)}{\Gamma(a)\Gamma(b)}\left(x + \frac{1}{2}\right)^{a-1}\left(\frac{1}{2} - x\right)^{b-1},$$

$a, b > 0$, and that of a shifted STSP distribution with the same unit support $[-0.5, 0.5]$ and the pdf

$$f(x|\theta, n) = \begin{cases} n\left(\dfrac{x+\frac{1}{2}}{m+\frac{1}{2}}\right)^{n-1}, & \text{for } -\frac{1}{2} \leq x < m \\[2ex] n\left(\dfrac{\frac{1}{2}-x}{\frac{1}{2}-m}\right)^{n-1}, & \text{for } m \leq x \leq \frac{1}{2}, \end{cases} \tag{3.6}$$

where $-0.5 \leq m \leq 0.5$, $n > 0$. The parameters of the beta distribution in Fig. 3.3 were fitted via a ML procedure (see, e.g., Mielke (1975)). The parameters of the shifted STSP distribution were fitted via a ML procedure developed in Van Dorp and Kotz (2002a) to be discussed in Sec. 3.3.

From Figs. 3.2 and 3.3 we observe that both the asymmetric Laplace and STSP pdf's provide a "superior" fit to the empirical cdf than the normal or beta density functions. A more formal fit analysis is conducted in Table 3.2. In Table 3.2 the chi-square statistic is calculated utilizing 15 bins. (Note that, $15 \in [\sqrt{196}, 196/5]$ as suggested by Banks *et al.* (2001)). The boundaries of the bins are selected such that the number of observations O_i, $i = 1, \ldots, 15$, in each Bin i equals 13 or 14, totaling 196 data points. Such a boundary selection procedure partitions the support of the observed data range in a similar manner as the "equal-probability method of constructing classes" (see, e.g., Stuart and Ord (1994)) while keeping the bin boundaries of the chi-square statistic

$$\sum_{i=1}^{15} \frac{(O_i - E_i)^2}{E_i} \tag{3.7}$$

the same across the five different distributions depicted in Table 3.2. The corresponding values E_i, $i = 1, \ldots, 15$, in (3.7) for the expected number of observations in Bin i are obtained using the formula

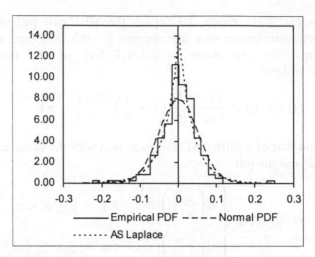

Fig. 3.2 Empirical pdf of two-step log differences of 30-year conventional mortgage interest rates together with the fitted Gaussian and asymmetric Laplace (Eq. (3.5)) distributions (using the ML method) ; Gaussian pdf ($\hat{\mu} = 1.067e - 3$, $\hat{\sigma} = 5.042e - 2$); Asymmetric Laplace pdf ($\hat{\mu} = (2.800e - 3$, $\hat{\sigma} = 5.104e - 2$, $\hat{\kappa} = 1.024$).

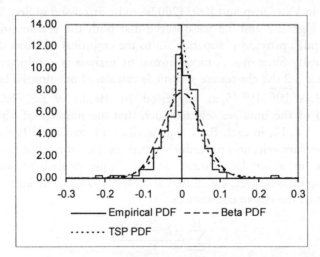

Fig. 3.3 Empirical pdf of two-step log differences of 30-year conventional mortgage interest rates together with fitted beta and TSP (Eq. (3.5)) distributions (using the ML method); Beta pdf ($\hat{a} = 47.424$, $\hat{b} = 47.229$); TSP pdf ($\hat{m} = 2.800e - 3$, $\hat{n} = 12.754$).
(Note that $\hat{m}$ is almost zero)

$$E_i = 196\{F(UB_i|\widehat{\Theta}) - F(LB_i|\widehat{\Theta})\},$$

where $F(\cdot|\widehat{\Theta})$ is the theoretical cdf, $\widehat{\Theta}$ are the corresponding ML estimators for each distribution and the bin boundaries $(LB_i, UB_i]$ are provided in Table 3.2. In addition, Table 3.2 presents the values of the Kolmogorov-Smirnov (KS) statistic D (see, e.g., Stuart and Ord (1994))

$$D = Max\{D_i|\, i = 1, \ldots, 196\} \tag{3.8}$$

where

Table 3.2 Goodness of fit analysis of ML fitted distributions for the 1966-2002 data on two-step (or bi-monthly) log-differences of monthly US CD interest rates.

Bin	LB_i	UB_i	O_i	Normal $(Oi\text{-}Ei)^2/Ei$	AS Laplace $(Oi\text{-}Ei)^2/Ei$	Beta $(Oi\text{-}Ei)^2/Ei$	TSP $(Oi\text{-}Ei)^2/Ei$
1	< -0.5	-0.0699	13	0.44	0.06	0.64	0.02
2	-0.070	-0.048	13	0.71	0.35	0.78	0.17
3	-0.048	-0.031	13	2.07	0.27	2.04	0.39
4	-0.031	-0.021	13	0.01	0.02	0.00	0.02
5	-0.021	-0.014	13	0.39	0.10	0.47	0.15
6	-0.014	-0.008	13	3.05	0.87	3.29	1.16
7	-0.008	-0.004	13	5.58	1.24	5.92	1.73
8	-0.004	0.006	14	0.16	5.15	0.11	3.90
9	0.006	0.012	13	2.38	0.01	2.59	0.09
10	0.012	0.020	12	0.01	0.82	0.00	0.65
11	0.020	0.028	13	0.61	0.29	0.69	0.27
12	0.028	0.037	13	0.02	0.19	0.03	0.10
13	0.037	0.046	13	1.07	3.37	1.09	2.57
14	0.046	0.065	13	0.63	0.20	0.68	0.01
15	0.065	> 0.5	14	1.86	0.33	2.23	0.80
Chi-Squared Statistic				18.99	13.27	20.56	12.01
Degrees of Freedom				12	11	12	12
P-value				0.089	0.276	0.057	0.445
K-S Statistic				8.90%	6.44%	9.21%	4.66%
SS				0.276	0.091	0.320	0.062
Log-Likelihood				307.42	319.16	304.71	318.36

$$D_i = Max\left\{|\frac{i-1}{196} - F(X_{(i)}|\widehat{\Theta})|, |\frac{i}{196} - F(X_{(i)}|\widehat{\Theta})|\right\},$$

and $X_{(i)}$, $i = 1, \ldots, 196$, are the order statistics associated with the data in Fig. 3.1C. Finally, values for an intuitive ad hoc measure of fit

$$\sum_{i=1}^{196}\left\{\frac{i}{196} - F(X_{(i)}|\widehat{\Theta})\right\}^2, \tag{3.9}$$

denoted by Sum of Squares (SS) (reminiscent of the popular sum of squares in linear regression analysis), and the log-likelihood

$$\sum_{i=1}^{196}Ln\{f(X_{(i)}|\widehat{\Theta})\} \tag{3.10}$$

are all provided in Table 3.2.

We observe from Table 3.2 that the beta and the normal distributions evidently produce a worst fit in terms of the chi-square statistic (3.7) as compared with the asymmetric Laplace and STSP distributions. In particular bins 6, 7, 8 and 9 (containing the peak of the empirical pdf) for the Gaussian and beta cases reconfirm the observation from Figs. 3.2 and 3.3 that the Gaussian and beta distributions do not adequately represent the "peak" in the data given in Fig. 3.1C. More importantly, the STSP distribution yields a substantial larger p-value (0.445) of the chi-squared hypothesis test than any one of the other three distributions in Table 3.2 when taking into account the number of parameters of each distribution to determine the degrees of freedom. While the asymmetric Laplace distribution (3.5) has one additional parameter as compared to the shifted STSP distribution (3.6), it has only a slight advantage over the shifted STSP distribution in terms of the log-likelihood statistic (3.10), but not in terms of the chi-squared statistic (3.7) (and its p-value), the KS-statistic (3.8) and the SS-statistic (3.9). In the authors' opinion the analysis in Figs 3.2 and 3.3, and Table 3.2 justify the use of the STSP distribution for the data of the type displayed in Fig. 3.1C. To the best of our knowledge the two-parameter STSP family was mentioned for the first time in passing in Nadarajah (1999) and appeared in Schmeiser and Lal (1985) as a special case of a five-parameter family (in a reparameterized form).

3.2 The Standard Two-Sided Power Distributions

After an informal but hopefully convincing motivation of the STSP distribution we shall now proceed with an organized description of it. Let X be a random variable with standard support $[0, 1]$ and the density function given by

$$
f(x|\theta, n) = \begin{cases} n\left(\dfrac{x}{\theta}\right)^{n-1}, & \text{for } 0 \leq x \leq \theta \\[3mm] n\left(\dfrac{1-x}{1-\theta}\right)^{n-1}, & \text{for } \theta \leq x \leq 1. \end{cases} \tag{3.11}
$$

X is said to follow a *Standard Two-Sided Power distribution*, $STSP\ (\theta, n)$, $0 \leq \theta \leq 1$, and $n > 0$, not necessarily an integer. Note that the pdf (3.6) is a shifted STSP distribution with support $[-0.5, 0.5]$ and with $m = \theta - \frac{1}{2}$.

For $0 \leq \theta \leq 1$ and $n > 0$, the density in (3.11) is unimodal with the mode at θ. For $0 < \theta < 1$ and $0 < n < 1$, the form of the density function in (3.11) take U-shaped forms with mode at 0 or 1. Figure 3.4 provides some examples of $STSP(\theta, n)$ distributions. For $n = 1$, the density given by (3.11) simplifies to the uniform$|0, 1|$ density (Fig. 3.4A), corresponds to a triangular density on $[0, 1]$ for $n = 2$ (Fig. 3.4B), and to a power distribution (Fig. 3.4C) for $\theta = 1$. Fig 3.4D displays the *reflected* power distribution corresponding to $\theta = 0$ with $n = 3$ and is the reflected version of Fig. 3.4C which corresponds to $\theta = 1$. The pdf's plotted in Figs. 3.4A, 3.4C and 3.4D are the common members to both the STSP and the beta families of distributions. Figures 3.4E to 3.4H, with a mode or an anti-mode strictly within $[0, 1]$, display members neither covered by the beta nor by the triangular family of distributions. Figure 3.4E displays the symmetric density by concatenating the power densities in Figs. 3.4C and 3.4D (and rescaled to $[0, 1]$) and motivates the designation "Two-Sided Power" family of distributions. A peaked generalization $(n = 6)$ of the triangular distribution $(n = 2)$ in Fig. 3.4B is presented in Fig. 3.4F. A "stumped" generalization $(n = 1.5)$ is displayed in Fig. 3.4G. Finally, a U-shaped form $(n = 0.5)$ similar to those occurring in the beta family is plotted in Fig. 3.4H. Note that, in Figs. 3.4A to 3.4G the modal value of the pdf equals the parameter n, whereas in Fig. 3.4H this parameter yields the value of the anti-mode.

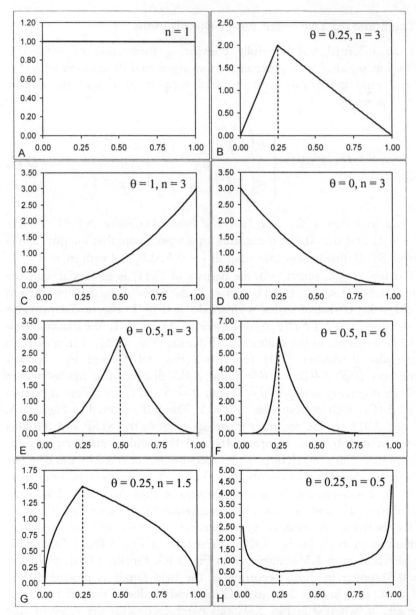

Fig 3.4 Examples of $STSP(\theta, n)$ distributions.

From (3.11) we obtain the cdf of a $STSP(\theta, n)$ distribution

$$F_{\theta,n}(x) = \begin{cases} \theta\left(\frac{x}{\theta}\right)^n, & \text{for } 0 \le x \le \theta \\ 1 - (1 - \theta)\left(\frac{1-x}{1-\theta}\right)^n, & \text{for } \theta \le x \le 1. \end{cases} \tag{3.12}$$

The reader is encouraged to examine cdf's on his/her own for particular values of n and θ to achieve familiarity with STSP distributions. In Sec. 3.2.2 we shall further discuss some properties of the cdf (3.12).

3.2.1 Moments

The k-th moment of a $STSP(\theta, n)$ derived from (3.11) is given by

$$E[X^k] = \frac{n\theta^{k+1}}{n+k} - \sum_{i=0}^{k} \binom{k}{k-i} \frac{n(\theta-1)^{i+1}}{n+i}. \tag{3.13}$$

Here calculations are a bit lengthy but straightforward. Substituting $k = 1$ we arrive at the first moment about zero

$$E[X] = \frac{(n-1)\theta + 1}{n+1} = 0\frac{1}{n+1} + \frac{n-1}{n+1}\theta + \frac{1}{n+1} \tag{3.14}$$

and substituting $k = 2$ yields the second moment about zero

$$E[X^2] = \frac{-2 \cdot (n-1)\theta^3 + 4(n-1)\theta^2 + (2 - 3n)\theta + n}{(n+2)(n+1)^2(1-\theta)}. \tag{3.15}$$

Utilizing (3.14), (3.15) and the relation $Var(X) = E[X^2] - E^2[X]$ we have

$$Var(X) = \frac{n - 2(n-1)\theta(1-\theta)}{(n+2)(n+1)^2}. \tag{3.16}$$

For a fixed value of n the variance attains its minimum at $\theta - \frac{1}{2}$ and its maximum at $\theta = 0$ or $\theta = 1$. These values are

$$\frac{1}{2(n+2)(n+1)}, \frac{n}{(n+2)(n+1)^2},$$

respectively.

The manner in which the right hand side (rhs) of Eq. (3.14) is written allows us to recognize the formula for the mean of the triangular distribution $(n = 2)$ on $[0, 1]$ as a *simple* average of the lower bound 0, the mode θ, and the upper bound 1, from which the triangular distribution derives its intuitive appeal (see, e.g., Williams (1992)). For the STSP distribution the mean is a weighted average of the lower bound 0, the location parameter θ, and the upper bound 1, where the weights are determined solely by the shape parameter n. For $n = 1$ (3.14) simplifies to $\frac{1}{2}$, the mean of a uniform$[0, 1]$ variable. For $n > 1$ more weight is assigned to the mode θ and less to the lower and upper bounds 0 and 1, respectively. This becomes more pronounced as the values of n increases. In the extreme case, $n \to \infty$, no weight is assigned to the bounds and the mean reduces to θ. For $n < 1$, the mean is discounted by θ at an increasing rate as n decreases while shifting the weight to the bounds. In the extreme case, $n \downarrow 0$, the mean simplifies to $1 - \theta$. In Sec. 3.2.4 a discussion of the limiting distributions is presented.

A simple relationships between the mean of a $STSP(\theta, n)$ distribution and its parameters renders this family to exhibit intuitive transparency similar to of the triangular distribution (see, e.g., D. Johnson (1997) and Williams (1992)).

3.2.2 Properties of the cdf

Alongside with the beta distribution, the $STSP$ distribution satisfies stochastically increasing and decreasing properties which means that "distributional parameters are working in a transparent manner". More precisely:

Theorem 3.1: *A. The cdf given by (3.12) is stochastically increasing for* $n > 1$, *i.e.*

$$\theta_1 < \theta_2, \; x \in (0, 1) \Rightarrow F(x|\theta_1, n) > F(x|\theta_2, n).$$

B. The cdf given by (3.12) is stochastically decreasing for $0 < n < 1$, *i.e.*

$$\theta_1 < \theta_2, \ x \in (0,1) \Rightarrow F(x|\theta_1, n) < F(x|\theta_2, n).$$

Proof: We shall prove statement A only. (For B the proof is analogous.) Let $0 < \theta_1 < \theta_2 < 1$, $n > 1$, $x \in (0,1)$. Three cases should be considered

$$(a)\, 0 < x < \theta_1, \ (b)\, \theta_2 < x < 1 \text{ and } (c)\, \theta_1 < x < \theta_2.$$

Cases (a) and (b) are straightforward since we are dealing here with a single branch of the distribution function (3.12). Consider Case (c). From (3.12) it follows that

$$F(x|\,\theta_1, n) = 1 - \frac{(1-x)^n}{(1-\theta_1)^{n-1}}, \ \ F(x|\,\theta_2, n) = \frac{x^n}{\theta_2^{n-1}}.$$

For $0 < \theta_1 < \theta_2 < 1$ one has the implications

$$\theta_1 < x < \theta_2 \Leftrightarrow \begin{cases} \frac{\theta_1}{\theta_2} < \frac{x}{\theta_2} < 1 \\ 1 > \frac{1-x}{1-\theta_1} > \frac{1-\theta_2}{1-\theta_1} \, . \end{cases} \tag{3.17}$$

From (3.17), the condition that $\theta_1 < x < \theta_2$, $n > 1$ and $x \in (0,1)$ we easily arrive at

$$(1-x)\left(\frac{1-x}{1-\theta_1}\right)^{n-1} < 1 - x < 1 - x\left(\frac{x}{\theta_2}\right)^{n-1}.$$

Hence

$$\frac{(1-x)^n}{(1-\theta_1)^{n-1}} < 1 - \frac{x^n}{\theta_2^n\,^1} \Leftrightarrow 1 - \frac{(1-x)^n}{(1-\theta_1)^{n-1}} > \frac{x^n}{\theta_2^{n-1}},$$

which proves statement A in this case. $\qquad\qquad\qquad\square$

3.2.3 Quantile properties

Denote by x_p the p-th percentile (also referred to as quantile) of (3.11) or (3.12), i.e. $F(x_p|\theta, n) = p$. The quantiles x_p enjoy the following properties:

Property 1 : $x_p < p \Leftrightarrow p < \theta.$

Property 2 :   Analogously, $x_p = \theta \Leftrightarrow p = \theta$, independently of the value of n.

Property 3 :   Consider values of p such that $p < Min(\theta, 1 - \theta)$. From (3.12) and Property 1, dealing separately with the two-part definition of $F(\cdot | \theta, n)$ (3.12), we have

$$\frac{x_p^n}{\theta^{n-1}} = \frac{(1 - x_{1-p})^n}{(1 - \theta)^{n-1}} \Leftrightarrow \tag{3.18}$$

$$\frac{x_p}{1 - x_{1-p}} = \left(\frac{\theta}{1 - \theta}\right)^{\frac{n-1}{n}}$$

       Hence the ratio $x_p/(1 - x_{1-p})$ does not depend on p for $p < Min(\theta, 1 - \theta)$.

Property 4 :   Analogously, for $p > Max(\theta, 1 - \theta)$,

$$\frac{x_{1-p}}{1 - x_p} = \left(\frac{\theta}{1 - \theta}\right)^{\frac{n-1}{n}} \tag{3.19}$$

       is independent of p.

From Property 2 it follows that for all STSP distributions the probability mass is split at θ into θ and $(1 - \theta)$ and hence will be referred to as the *hinge* property. In other words, the probability mass to the left (right) of θ equals the relative distance of the mode θ to its lower bound (upper bound) over the support $[0, 1]$ regardless of the value of n. Properties $2, 3$ and 4 may be utilized for an initial estimation of the parameter θ. Properties 3 and 4 involve non-trivial relations satisfied by the complementing quantiles x_p and x_{1-p}. The properties above are of course valid for triangular distributions on $[0, 1]$ as well.

3.2.4 Some limiting distributions

Let $X \sim STSP(\theta, n)$. Setting $\theta = 0$ and letting $n \to \infty$ it follows from (3.14) and (3.16) that the distribution of X converges to a degenerate distribution with a point mass of 1 at 0. Analogously, setting $\theta = 1$ and $n \to \infty$, we observe that the distribution of X converges to a degenerate distribution of mass 1 at 1. Setting $0 < \theta < 1$, and letting $n \to \infty$ it follows that the distribution of X converges to a degenerate distribution with a point mass of 1 at θ. As $n \downarrow 0$ and $0 < \theta < 1$, we obtain from (3.13) that $E[X^k] \to (1 - \theta)$, for all k. Hence, when $n \downarrow 0$, $E[X^k]$, $k = 1, 2, \dots$, converge to the moments of a Bernoulli variable with a point mass of $1 - \theta$ at 1. Since both X and a Bernoulli variable have bounded support, we conclude from the uniqueness theorem for distributions with a bounded support that the Bernoulli distribution with a point mass of $1 - \theta$ at 1 is the limiting distribution of X for $0 < \theta < 1$ as $n \downarrow 0$ (see, e.g., Harris (1966) p. 103 for a discussion of the uniqueness theorem). These limiting distributions coincide with related limiting distributions of the beta distribution (see, e.g., van Dorp and Mazzuchi (2000)). In other words, the flexibility of the $STSP(\theta, n)$ class is on par with that of the beta family.

Another interesting limiting distribution (associated with the leading example in Sec. 5.1) can be derived using the linear transformation

$$Y = \frac{(n-1)A}{\theta}(X - \theta), \tag{3.20}$$

$A > 0$ being an arbitrary positive constant. From (3.11), (3.20) and $n > 1$ it follows that

$$f(y|\theta, n, A) =$$

$$\begin{cases} \frac{n\theta}{(n-1)A}\left(1 + \frac{1}{(n-1)}\frac{y}{A}\right)^{n-1}, & \text{for } (1-n)A \le y \le 0 \\ \frac{n\theta}{(n-1)A}\left(1 - \frac{1}{(n-1)}\frac{\theta y}{(1-\theta)A}\right)^{n-1}, & \text{for } 0 \le y \le \frac{(1-\theta)(n-1)A}{\theta}. \end{cases}$$

Letting $n \to \infty$, we have

$$f(y|\theta, n, A) \to f(y|\theta, A) = \begin{cases} \frac{\theta}{A}exp\left(\frac{y}{A}\right), & \text{for } y \le 0 \\ \frac{\theta}{A}exp\left(-\frac{\theta}{1-\theta}\frac{y}{A}\right), & \text{for } y > 0. \end{cases} \tag{3.21}$$

Hence, $f(y|\theta, n, A)$ converges to the density $f(y|\theta, A)$ of an "asymmetric Laplace" variable (see, e.g., Johnson *et al.* (1994)) where its probability mass is split at 0 into θ and $(1 - \theta)$, regardless of the value of the parameter A.

The asymmetric Laplace distribution (3.5) considered by Kozubowski and Podgorski (1999) simplifies to (3.21) by reparameterizing

$$\kappa = \sqrt{\frac{1-\theta}{\theta}}, \sigma = A\sqrt{\frac{2(1-\theta)}{\theta}}.$$

and setting $\mu = 0$.

3.2.5 Relative entropy

The relative entropy (also known as cross entropy or discrimination function) of an absolute continuous pdf $f(x|\Theta)$ with respect to a probability mass function $g(x)$ is defined as

$$E(f : g|\Theta) = \int Log\frac{f(x|\Theta)}{g(x)}dF(y|\Theta), \qquad (3.22)$$

and is used as a measure for comparing the information contents of distributions. The term *discrimination* reflects that $E(f; g|\Theta) \geq 0$ and the equality holds if and only if $f(x|\Theta) = g(x)$ almost everywhere (see, e.g., Soofi and Retzer (2000)). We compare the information content of STSP distributions on $[0, 1|$ with the information content of an uniform$|0, 1|$ distribution. The relative entropy of beta distributions with respect to a uniform$|0, 1|$ distribution has been found to be (see, e.g., Soofi and Retzer (2000))

$$E(f : g|a, b) = Log(\mathbb{B}(a, b) - \qquad (3.23)$$
$$(a-1)(\psi(a) - \psi(a+b)) - (b-1)(\psi(b) - \psi(a+b))$$

where

$$f(x|a, b) = \frac{1}{\mathbb{B}(a, b)} x^{a-1}(1-x)^{b-1}, \qquad (3.24)$$

$a > 0$, $b > 0$, $\mathbb{B}(a, b) = \frac{\Gamma(a)\Gamma(b)}{\Gamma(a+b)}$, g is the uniform$[0, 1]$ pdf and $\psi(\,\cdot\,) = \Gamma'(\,\cdot\,)$ is the psi-function.

    The relative entropy of STSP distributions with respect to a uniform$|0, 1|$ distribution, using the definition (3.22), results in

$$E(f : g \,|\, \theta, n) = Log\, n - \frac{n-1}{n},$$ (3.25)

where f is the STSP density given by (3.11) and g is as above. Note that it follows formally from (3.25) that $E(f : g | \theta, n)$ attains its minimal value 0 when the STSP variable coincides with a uniform$|0, 1|$ variable, i.e. $n = 1$. This, of course, is evident from the verbal definition of relative entropy. Note also that the forms of (3.25) and (3.23) are similar, except that $E(f : g | \theta, n)$ is constant for fixed n regardless of the value θ. Hence, no information is added to or subtracted from the information content of a STSP distribution by varying the parameter θ, while keeping n fixed (this property is perhaps somewhat counter-intuitive). Consequently, the relative entropy of all triangular distributions f on $[0, 1]$ equals $-\frac{1}{2} + Log(2) \approx 1.193$, regardless of the location of the mode θ, where g is a uniform distribution on $[0, 1]$. It is worth noting that the variance (which is intuitively related to entropy) of an $STSP(\theta, n)$ variable <u>does</u> depend on θ. (See the discussion following Eq. (3.16).)

3.3 Maximum Likelihood Method of Estimating Parameters

The structure of the STSP distribution (3.11) leads to an rather ingenious procedure for estimating its parameters, especially the threshold parameter θ. This parameter can be viewed as "dividing" (in the sense that it is related to two different analytical expressions appearing in the definition of the STSP density). Roughly speaking, the smaller sample observations obey a different probability law than the larger ones. The same phenomenon could be traced to the triangular distribution in Chapter 1.

The derivation of the ML estimation procedure of a STSP distribution (3.11) to be described below is quite instructive (and is similar to the ML procedure for the triangular distribution discussed in Sec. 1.4 in Chapter 1). The reader is advised to review the proof of Theorem 1.1 in Chapter 1. Let for a random i.i.d. sample of size s, $\underline{X} = (X_1, \ldots, X_s)$, the order statistics be $X_{(1)} < X_{(2)} < \ldots < X_{(s)}$. By definition, the likelihood for X with distribution (3.11) is

$$L(\underline{X} ; \theta, n) = n^s \{H(\underline{X}; \theta)\}^{n-1}$$ (3.26)

where

$$H(\underline{X};\theta) = \frac{\prod\limits_{i=1}^{r} X_{(i)} \prod\limits_{i=r+1}^{s} (1 - X_{(i)})}{\theta^r(1 - \theta)^{s-r}} \tag{3.27}$$

and $X_{(r)} \leq \theta \leq X_{(r+1)}$ with $X_{(0)} \equiv 0$, $X_{(s+1)} \equiv 1$. We shall first consider a ML procedure for the parameter domain $n \geq 1$ and $0 \leq \theta \leq 1$ involving the unimodal and uniform members of the pdf (3.11). This means that one should be able *a priori* to rule out the U-shaped case (i.e. when $0 < n < 1$) for the data under consideration (via, for example, a histogram of the data). We shall discuss further details for the U-shaped case of the pdf after proving the theorem below[1].

Theorem 3.2: *Let* $\underline{X} = (X_1, \ldots, X_s)$ *be an i.i.d. sample from a* $STSP(\theta, n)$ *distribution. The ML estimators of* θ *and* n *maximizing the likelihood (3.26) over the parameter domain* $0 \leq \theta \leq 1$ *and* $n > 1$ *are:*

$$\begin{cases} \widehat{\theta} = X_{(\widehat{r})} \\ \widehat{n} = Max\left\{ -\frac{s}{Log\, M(\widehat{r})}, 1 \right\}, \end{cases} \tag{3.28}$$

where

$$\widehat{r} = \underset{r \in \{1, \ldots, s\}}{arg\, max} M(r) \tag{3.29}$$

and

$$M(r) = \prod_{i=1}^{r-1} \frac{X_{(i)}}{X_{(r)}} \prod_{i=r+1}^{s} \frac{1 - X_{(i)}}{1 - X_{(r)}}. \tag{3.30}$$

Proof: To maximize the likelihood (3.26), we represent it as

---

[1]The proof of Theorem 3.2 is based on the exposition provided in our 2002 paper "The standard two sided power distribution and its properties: with applications in financial engineering" (*The American Statistician*, 56 (2), pp. 90-99). Unfortunately, the discussion in the paper is incomplete. The proof in the paper claims that both cases (the unimodal and the U-shaped) yield the same ML estimator. A corrected proof is presented herein.

$$\underset{n \geq 1,\, 0 \leq \theta \leq 1}{max} L(\underline{X}; \theta, n) = \underset{n \geq 1}{max}\; n^s \widehat{M}^{n-1}, \qquad (3.31)$$

where $\widehat{M}$ is given by

$$\widehat{M} = \underset{0 \leq \theta \leq 1}{max} H(\underline{X}; \theta), \qquad (3.32)$$

$H(\underline{X}; \theta)$ is defined by (3.27) and, as above, $X_{(r)} \leq \theta \leq X_{(r+1)}$, with $X_{(0)} \equiv 0$, $X_{(s+1)} \equiv 1$. Now

$$Log\left(n^s \widehat{M}^{n-1}\right) = (n-1)Log\,\widehat{M} + s\,Log\,n, \qquad (3.33)$$

and

$$\frac{\partial}{\partial n}Log\left(n^s \widehat{M}^{n-1}\right) = Log\,\widehat{M} + \frac{s}{n}. \qquad (3.34)$$

Equating the partial derivative (3.34) to zero yields

$$n^* = -(s/Log\,\widehat{M}). \qquad (3.35)$$

From (3.34) it follows that

$$\frac{\partial}{\partial n}Log\left(n^s \widehat{M}^{n-1}\right) > 0 \Leftrightarrow n^* < -\frac{s}{Log\,\widehat{M}}.$$

Hence n^* corresponds to a global maximum of (3.33) and the likelihood (3.31). Note that, for $i < r$, we have $0 < X_{(i)}/\theta < 1$ and for $i > r$, $0 < (1 - X_{(i)})/(1 - \theta) < 1$. Hence $0 < \widehat{M} < 1$ and thus $n^* > 0$. From the restriction $n \geq 1$ and the fact that $n^* > 0$ is a global maximum it follows from (3.35) that

$$\widehat{n} = Max\left\{-(s/Log\,\widehat{M}), 1\right\} \qquad (3.36)$$

is the ML estimator for n. The ML estimator $\widehat{\theta} = X_{(\widehat{r})}$ in (3.28) follows immediately from the fact that the function $H(\underline{X}; \theta)$ in (3.27) is identical to Eq. (1.25) in Chapter 1 and the proof of Theorem 1.1 in Chapter 1 can be applied. In addition, in Theorem 1.1 it has been shown that the value $\widehat{M}$ in

(3.32) equals $M(\widehat{r})$, where $\widehat{r}$ is given by (3.29) and $M(r)$ is defined by (3.30). $\qquad\qquad\qquad\qquad\qquad\qquad\qquad\qquad\qquad\qquad\qquad\qquad$ $\square$

We shall now consider some modifications required for the case of U-shaped STSP distributions (3.11) associated with the parameter domain $0 < n < 1, 0 \leq \theta \leq 1$. We have (similar to Eq. (3.31)):

$$\underset{0 < n < 1,\, 0 \leq \theta \leq 1}{max} L(\underline{X}; \theta, n) = \underset{0 < n < 1}{max} \; n^s \, \widetilde{m}^{\,n-1}, \quad (3.37)$$

where

$$\widetilde{m} = \underset{0 \leq \theta \leq 1}{min} H(\underline{X}; \theta) \qquad\qquad (3.38)$$

(compare with (3.32)) and $H(\underline{X}; \theta)$ as defined by Eq. (3.27). Analogously to Eq. (1.32) and utilizing (3.38) one obtains the global minimum $\widetilde{m}$ of $H(\underline{X}; \theta)$ via $s + 1$ minimizations over the sets $[X_{(r)}, X_{(r+1)}]$. Specifically,

$$\widetilde{m} = \underset{r \in \{0, \dots, s\}}{min} h(r), \qquad\qquad (3.39)$$

where

$$h(r) = \underset{X_{(r)} \leq \theta \leq X_{(r+1)}}{min} H(\underline{X}; \theta), \qquad\qquad (3.40)$$

$r = 0, \dots, s$, $X_{(0)} \equiv 0$ and $X_{(s+1)} \equiv 1$ (compare with (1.32)). For the cases $r = 0$ and $r = s$, we obtain from (3.40) and the definition of $H(\underline{X}; \theta)$ (3.27):

$$h(r) = \begin{cases} H(\underline{X}; 0) = \prod\limits_{i=1}^{s} 1 - X_{(i)}, & \text{for } r = 0, \\[2mm] H(\underline{X}; 1) = \prod\limits_{i=1}^{s} X_{(i)}, & \text{for } r = s, \end{cases} \qquad (3.41)$$

since now we are required to minimize $H(\underline{X}; \theta)$ (compare with Eqs. (1.37) and (1.39), respectively). For the cases $r = 1, \dots, s - 1$ of (3.40), it follows that here one needs to include the unique stationary point $\theta^* = r/s$ of the function $g(\theta)$ (Eq. (1.33) in Chapter 1) when minimizing $H(\underline{X}; \theta)$ over these sets $[X_{(r)}, X_{(r+1)}]$. Hence we obtain, utilizing (3.41):

$$h(r) = \begin{cases} M(r), & \text{if } \frac{r}{s} < X_{(r)}, \\ H(\underline{X};\frac{r}{s}), & \text{if } X_{(r)} \leq \frac{r}{s} \leq X_{(r+1)}, \\ M(r+1), & \text{if } \frac{r}{s} > X_{(r+1)}, \end{cases} \tag{3.42}$$

for $r = 0, \ldots, s$, where the product $M(r)$ is given by (3.30). Thus, from (3.37), (3.38), (3.39) and (3.42) the ML estimator of the parameter θ for the U-shaped case of the STSP distribution (3.11) (i.e. $0 < n < 1$) becomes

$$\widetilde{\theta} = \begin{cases} X_{(\widetilde{r})}, & \text{if } \frac{\widetilde{r}}{s} < X_{(\widetilde{r})}, \\ \frac{\widetilde{r}}{s}, & \text{if } X_{(\widetilde{r})} \leq \frac{\widetilde{r}}{s} \leq X_{(\widetilde{r}+1)}, \\ X_{(\widetilde{r}+1)}, & \text{if } \frac{\widetilde{r}}{s} > X_{(\widetilde{r}+1)}, \end{cases} \tag{3.43}$$

where

$$\widetilde{r} = \underset{r \in \{0,\ldots,s\}}{\arg\min} \; h(r). \tag{3.44}$$

(Compare with (3.29)) and $h(r)$ is defined by (3.42).) Analogously to (3.35), we have the ML of the parameter n over the domain $0 < n < 1$ to be

$$\widetilde{n} = Min\left\{ -(s/Log\,\widetilde{m}), 1 \right\}, \tag{3.45}$$

where $\widetilde{m}$ is defined by (3.39). Hence in the U-shaped case of the STSP distribution (3.11) it is possible that the ML estimator $\widetilde{\theta}$ is *not* being attained at one of the order statistics $X_{(r)}$. (Compare with $\widehat{\theta}$ in (3.28).)

To maximize the likelihood $L(\underline{X};\theta,n)$ (3.26) over the whole parameter domain $0 \leq \theta \leq 1$, $n > 0$, one compares the value of $L(\underline{X};\widehat{\theta},\widehat{n})$ to that of $L(\underline{X};\widetilde{\theta},\widetilde{n})$. If

$$L(\underline{X};\widehat{\theta},\widehat{n}) > L(\underline{X};\widetilde{\theta},\widetilde{n})$$

the ML estimators for θ and n (over the entire parameter domain) are $\widehat{\theta}$ and $\widehat{n}$ as given by (3.28). Otherwise, the ML estimators for θ and n are $\widetilde{\theta}$, $\widetilde{n}$ as given by (3.43) and (3.45), respectively. □

3.3.1 An illustrative example

We shall illustrate the ML estimation procedure for an $STSP(\theta, n)$ distribution by means of the following hypothetical order statistics

$$(X_{(1)}, \ldots, X_{(8)}) = (0.10, 0.25, 0.30, 0.40, 0.45, 0.60, 0.75, 0.80), \quad (3.46)$$

which were also used in Chapter 1 (Eq. (1.41)) in connection with the ML method for triangular distributions. Figure 3.5 depicts the likelihood function $L(X ; \theta, n)$ (3.26) for the data in (3.46).

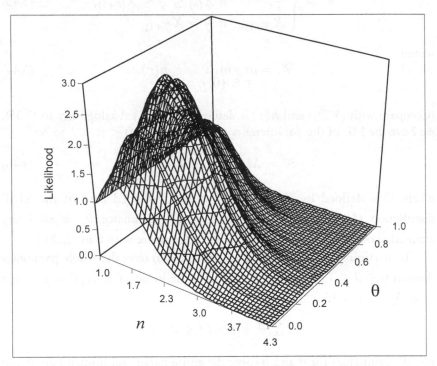

Fig. 3.5 Graph of the likelihood $L(X ; \theta, n)$ (Eq. (3.26)) for the data in (3.46).

From Fig. 3.5 we may visually observe that $L(X ; \theta, n)$ attains a global maximum over the domain $n > 1$, hence we shall utilize the ML estimators $\widehat{\theta}$ and $\widehat{n}$ in (3.28) of Theorem 3.2. Consider the matrix $A = [a_{i,r}]$ with the entries defined by Eq. (1.43) in Chapter 1 and Table 1.1 which summarizes calculations of the matrix A for the order statistics given in (3.46). Recall that the last row in the Table 1.1 contains the products of the matrix entries in the r-th column which are equal to the values of $M(r)$ given by (3.30) (and Eq. (1.28) in Chapter 1), $r = 1, \ldots, s$. Here, $s = 8$. From the last row of Table 1.1 and utilizing Eqs. (1.40), (3.29) and (3.28) we have here

$$\widehat{M} = 0.011; \; \widehat{r} = 3;$$
$$\widehat{\theta} = X_{(\widehat{r})} = 0.30; \tag{3.47}$$
$$\widehat{n} = \frac{-8}{Log(\widehat{M})} = 1.771.$$

Figure 1.5 in Chapter 1 displays the function $H(\underline{X};\theta)$ defined by Eq. (3.27) and shows that for the data in (3.46) the maximum value $\widehat{M} = 0.011$ of $H(\underline{X};\theta)$ over $\theta \in [0, 1]$ is attained at $X_{(3)} = 0.30$. Note also that the maximum value $H(r)$ (see, Eq. (1.15)) of $H(\underline{X};\theta)$ over $\theta \in [X_{(r)}, X_{(r+1)}]$ is attained at either $X_{(r)}$ or $X_{(r+1)}$ for all $r = 0, \ldots, s$.

Considering now the U-shaped domain $(0 < n < 1)$ for the example data (3.46), we conclude immediately from Fig. 1.5 in Chapter 1 utilizing (3.39) and the definition of $H(\underline{X};\theta)$ (3.27) that $\widetilde{m} = H(\underline{X};1) \approx 4.86 - 4$ (attained at $\widetilde{\theta} = 1$). Hence, from Eq. (3.45) we have

$$\widetilde{n} = Min\{1.04, 1\} = 1. \tag{3.48}$$

Next, from (3.26) it follows that $L(\underline{X};\widetilde{\theta}, \widetilde{n}) = 1$, whereas applying (3.47) it follows that $L(\underline{X};\widehat{\theta}, \widehat{n}) \approx 2.97$ (Compare with Fig. 3.5). Hence, the ML estimators $\widehat{\theta}, \widehat{n}$ given in (3.47) maximize the likelihood over the *whole* parameter domain $0 \le \theta \le 1, n > 0$.

ML estimation of the parameters of $X \sim STSP(\theta, n)$ can be modified to the two-parameter maximum likelihood estimation for $Z \sim TSP(a, m, b, n)$, where $Z = (b - a)X + a$. Here the parameters a and b are fixed and the parameter $m = (b - a)\theta + a$. The pdf of Z is

$$f(z|a, m, b, n) = \begin{cases} \frac{n}{(b-a)}\left(\frac{z-a}{m-a}\right)^{n-1} & a \le z \le m \\ \frac{n}{(b-a)}\left(\frac{b-z}{b-m}\right)^{n-1} & m \le z \le b. \end{cases} \tag{3.49}$$

The maximum likelihood estimators of the parameters m and n in the unimodal domain (i.e. when $n \ge 1$) of the distribution (3.49) utilizing order statistics $(Z_{(1)}, \ldots, Z_{(s)})$ are

$$\begin{cases} \widehat{m}(a,b) = Z_{(\widehat{r}(a,b))} \\ \widehat{n}(a,b) = Max\left\{1, \ -\frac{s}{Log\,M(a,b,\widehat{r}(a,b))}\right\}, \end{cases} \tag{3.50}$$

where, as above,

$$\widehat{r}(a,b) = \underset{r \in \{1, \ldots, s\}}{arg\ max}\ M(a,b,r) \tag{3.51}$$

and

$$M(a,b,r) = \prod_{i=1}^{r-1} \frac{Z_{(i)} - a}{Z_{(r)} - a} \prod_{i=r+1}^{s} \frac{b - Z_{(i)}}{b - Z_{(r)}}. \tag{3.52}$$

(Compare with (3.30) which corresponds to the case $b = 1$, $a = 0$.) The ML estimators in (3.50) were used in the leading example in Sec. 3.1, with fixed $a = -0.5$ and $b = 0.5$, yielding

$$\widehat{m}(-0.5, 0.5) = 2.800e - 3, \ \widehat{n}(-0.5, 0.5) = 12.754 \cdot \tag{3.53}$$

(The authors have also studied the four-parameter ML estimation for the $TSP(a,m,b,n)$ distribution (3.49), which will be discussed in the next chapter.)

3.4 Method of Moments

It was shown in Sec. 3.1 that the $TSP(a,m,b,n)$ distribution (3.49) with support $[a,b] = [-0.5, 0.5]$ and ML parameters (3.53) provides an adequate fit to the empirical density function displayed in Fig. 3.3 dealing with mortgage rates (one of the most common financial responsibilities in the majority of North-American households). Recall that the empirical pdf in Fig. 3.1C was constructed from bi-monthly log-differences of 30-year conventional mortgage rates. We shall illustrate the method of moments estimation of the parameters for a $STSP(\theta, n)$ distribution by again utilizing the bi-monthly log-differences of 30-year conventional mortgage rates. First, however, we shall shift the data with (the assumed) original support $[-0.5, 0.5]$ to the standard support $[0,1]$ of a $STSP(\theta, n)$ distribution. For the shifted data we have

$$\overline{x} = \frac{1}{2} + \frac{1}{196}\sum_{k=1}^{196} Ln(\epsilon_{k,2}) = 5.011e - 1, \qquad (3.54)$$

and

$$\widehat{\sigma}^2 = \frac{1}{195}\sum_{i=1}^{196}\{Ln(\epsilon_{k,2}) + 0.5 - \overline{x}\}^2 = 2.555e - 3, \qquad (3.55)$$

where $Ln(\epsilon_{k,2})$ is defined by (3.4). Equating the sample quantities $\overline{x}$, $\widehat{\sigma}^2$ with their population counterparts (3.14) and (3.16), respectively, we arrive at the following cubic equation in the parameter n

$$cn^3 + dn^2 + en + f = 0. \qquad (3.56)$$

The coefficients are given by

$$c = \frac{1}{2}\,\widehat{\sigma}^2, d = \widehat{\sigma}^2, \qquad (3.57)$$

$$e = -\{(\overline{x} - \frac{1}{2})^2 + \frac{1}{2}\,\widehat{\sigma}^2 + \frac{1}{4}\},$$

$$f = \frac{1}{4} - (\overline{x} - \frac{1}{2})^2 - \widehat{\sigma}^2.$$

After solving the cubic equation (3.56) for $\widehat{n}$ (see below), the estimate of θ follows from

$$\widehat{\theta} = \frac{1}{2} + \frac{\widehat{n} + 1}{\widehat{n} - 1}(\overline{x} - \frac{1}{2}). \qquad (3.58)$$

since the expression (3.14) was utilized to derive (3.56). The cubic equation (3.56) may be solved using a classical method known as Cardano's method. This method was discovered by Tartaglia in 1539 and later was published by Cardano in 1545 (see, Cardano (1993) — an English translation — for full details).

We shall briefly present a version of Cardano's method and demonstrate its application utilizing the values for $\overline{x}$ and $\widehat{\sigma}^2$ in (3.54) and (3.55), respectively. Dividing the LHS of (3.56) by c and introducing

$$y = n + \frac{d}{3c}, \tag{3.59}$$

we obtain the simplified cubic equation

$$y^3 + 3py + 2q = 0, \tag{3.60}$$

where

$$3p = \frac{3ce - d^2}{3c^2}, \quad 2q = \frac{2d^3}{27c^3} - \frac{de}{3c^2} + \frac{f}{c}. \tag{3.61}$$

Table 3.3 summarizes a solution method for solving the cubic equation (3.60).

Table 3.3 A version of Cardano's method for solving the simplified cubic equation (3.60).

$r = sign(q)\sqrt{|p|}$		
$p < 0$		$p > 0$
$q^2 + p^3 \leq 0$	$q^2 + p^3 > 0$	
$\cos\varphi = \frac{q}{r^3}$	$\cosh\varphi = \frac{q}{r^3}$	$\sinh\varphi = \frac{q}{r^3}$
$y_1 = -2r\cos\frac{\varphi}{3}$	$y_1 = -2r\cosh\frac{\varphi}{3}$	$y_1 = -2r\sinh\frac{\varphi}{3}$
$y_2 = 2r\cos(60° - \frac{\varphi}{3})$	$y_2 = r\cosh\frac{\varphi}{3} +$	$y_2 = r\sinh\frac{\varphi}{3} +$
	$i\sqrt{3}\,r\sinh\frac{\varphi}{3}$	$i\sqrt{3}\,r\cosh\frac{\varphi}{3}$
$y_2 = 2r\cos(60° + \frac{\varphi}{3})$	$y_3 = r\cosh\frac{\varphi}{3} -$	$y_3 = r\sinh\frac{\varphi}{3} -$
	$i\sqrt{3}\,r\sinh\frac{\varphi}{3}$	$i\sqrt{3}\,r\cosh\frac{\varphi}{3}$

Substituting the values of $\overline{x}$ and $\widehat{\sigma}^2$ given in (3.54) and (3.55) into (3.57) yields

$$c = 1.277e - 3, \tag{3.62}$$
$$d = 2.555e - 3,$$
$$e = -2.513e - 1,$$
$$f = 2.474e - 1.$$

Using the values for $c, d, e,$ and f in (3.62) and solving for p and q we obtain

$$p = -66.012,$$
$$q = 162.715.$$

Since $q^2 + p^3 = -261180.087$ (a negative number), the following three real-valued solutions of the cubic equation (3.60) are calculated utilizing the left most column of Table 3.3:

$$\widehat{y}_1 = -14.832, \; \widehat{y}_2 = 13.165, \; \widehat{y}_3 = 1.667.$$

Using the original values for c, d and (3.59) yields the following three solutions for the cubic equation (3.56):

$$\widehat{n}_1 = -15.498, \; \widehat{n}_2 = 12.498, \; \widehat{n}_3 = 1.000.$$

The first solution $\widehat{n}_1$ is inadmissible since we must have $n > 0$. Hence, we solve for $\widehat{\theta}_2$ and $\widehat{\theta}_3$ using $\widehat{n}_2$, $\widehat{n}_3$ and (3.58), yielding

$$\widehat{\theta}_2 = 0.501, \; \widehat{\theta}_3 = -226.524. \tag{3.63}$$

The third solution $\widehat{n}_3$ is now disqualified since we must have $0 \leq \theta \leq 1$. Using the relation $m = (b - a)\theta + a$ with $a = -0.5$ and $b = 0.5$ and $\widehat{\theta}_2$ as given in (3.63) the *unique* moment estimators for the data in Fig. 3.1C are

$$\widehat{m}_2 = 1.253e - 3, \; \widehat{n}_2 = 12.532. \tag{3.64}$$

Table 3.4 compares the ML fitted TSP distribution with parameters (3.53) to the method of moments fitted distribution with parameters (3.64), utilizing the same goodness-of-fit statistics as in Table 3.2.

Table 3.4 Fit analysis comparing ML fitted TSP distribution with parameters (3.53) and the method of moments (MM) fitted TSP distribution with parameters (3.64) for the data in Fig. 3.1C.

	TSP ML	TSP MM
Chi-Squared Statistic	12.01	12.25
Degrees of Freedom	12	12
P-value	0.445	0.426
K-S Statistic	4.66%	5.51%
SS	0.062	0.080
Log-Likelihood	318.36	318.17

Note that the methods of moments fitted TSP distribution performs slightly worse than the ML fitted one in terms of all the statistics in Table 3.2. Also note that of the criteria in Table 3.2 associated with the asymmetric Laplace distribution (with an additional parameter), only the log-likelihood criterion marginally outweighs the method of moments fitted TSP distribution with parameters (3.64).

3.5 Moment Ratio Diagram Comparison with the Beta Family

Moment ratio plots, popularized for Pearson-type distributions by Elderton and Johnson (1969), seem to provide a useful visual (graphical) assessment of the skewness (asymmetry) and the elusive kurtosis (peakedness) inherent in a particular family of asymmetric distributions. The classical form of the diagram shows the values of the ratios

$$\beta_1 = \frac{E^2[(X - E[X])^3]}{E^3[(X - E[X])^2]} = \frac{\mu_3^2}{\mu_2^3}, \ \beta_2 = \frac{E[(X - E[X])^4]}{E^2[(X - E[X])^2]} = \frac{\mu_4}{\mu_2^2},$$

where β_1 is plotted on the abscissa and β_2 on the ordinate. This diagram possesses a disadvantage that the sign of μ_3 (indicating left skewness or right skewness) disappears. A moment ratio diagram that retains this information is a plot with $\sqrt{\beta_1}$ on the abscissa and β_2 on the ordinate, with the convention that $\sqrt{\beta_1}$ retains the sign of μ_3 (See, e.g., Kotz and Johnson (1985)).

Values for $\sqrt{\beta_1}$ and β_2 for STSP distributions can be calculated using the general expression for the moments around the origin $\mu_k' = E[X^k]$, $k = 1, \ldots, 4$, (3.13) and their relationship with central moments μ_k, $k = 2, 3, 4$ given by

$$\begin{cases} \mu_2 = \mu_2' - {\mu_1'}^2 \\ \mu_3 = \mu_3' - 3\mu_2'\mu_1' + 2{\mu_1'}^3 \\ \mu_4 = \mu_4' - 4\mu_3'\mu_1' + 6\mu_2'{\mu_1'}^2 - 3{\mu_1'}^4 \end{cases}$$

(see, e.g., Stuart and Ord (1994)). Explicit forms of $\sqrt{\beta_1}$ and β_2 for STSP distributions result in cumbersome and not very informative expressions which are omitted. Figure 3.6 displays the moment ratio diagram coverage for the STSP family given by (3.11) restricted to a parameter range of

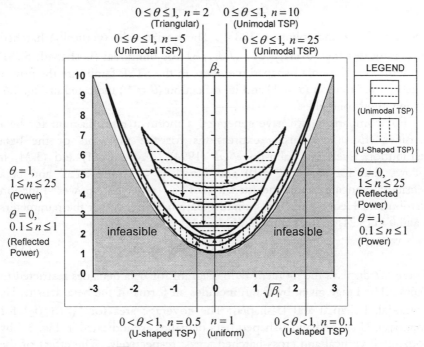

Fig. 3.6 ($\sqrt{\beta_1}, \beta_2$) Moment ratio diagram for STSP distributions
with parameters in the range $0 \le \theta \le 1, 0.1 \le n \le 25$.

$$0.1 \le n \le 25, 0 \le \theta \le 1. \qquad (3.65)$$

The range indicated by (3.65) is a plausible range for many practical purposes and includes unimodal forms $(0 \le \theta \le 1, n > 1)$, U-shaped forms $(0 \le \theta \le 1, n < 1)$, the uniform distribution $(n = 1)$, triangular distributions $(n = 2)$, J-shaped power function distributions $(\theta = 1, n > 0)$ and their J-shaped reflections $(\theta = 0, n > 0)$. Figure 3.6 also shows the effect of the parameters (θ, n) on $(\sqrt{\beta_1}, \beta_2)$ for specific examples of these cases indicated by solid lines in the moment ratio diagram. The shaded region in the moment ratio diagram is called the *infeasible* region since for all distributions we must have the relationship:

$$\beta_2 \geq (\sqrt{\beta_1})^2 + 1$$

(see, e.g., Kotz and Johnson (1985)). The horizontally (vertically) hatched areas indicates the coverage of $(\sqrt{\beta_1}, \beta_2)$ for unimodal (U-shaped) STSP distributions. The only J-shaped members in the STSP family are the power function distribution $(\theta = 1)$ and its reflection $(\theta = 0)$ indicated in Fig. 3.6. by only their solid lines.

For comparison, we have generated a moment ratio diagram for beta densities (Eq. (3.24)) using expressions for the moments of the beta distribution (see, e.g., Johnson *et al.* (1994)). From (3.11) and (3.24) it follows that for $\theta = 1$ $(\theta = 0)$ and $b = 1$ $(a = 1)$ the STSP and beta densities coincide with the density of a (reflected) power function distribution. Hence, the corresponding parameter range for the parameters a and b in (3.24) follows from (3.65) to be

$$0.1 \leq a \leq 25, \, 0.1 \leq b \leq 25. \tag{3.66}$$

Figure 3.7 displays the moment ratio diagram for beta densities restricted to (3.66). The range given by (3.66) includes all forms of the beta density, i.e. unimodal, J-shaped and U-shaped. The coverage area for $(\sqrt{\beta_1}, \beta_2)$ for unimodal, U-shaped and J-shaped beta densities are indicated in Fig. 3.7 by horizontal, vertical and cross-hatched areas, respectively. The effect of the parameters a and b in (3.24) on $(\sqrt{\beta_1}, \beta_2)$ is described by solid lines for those cases of beta densities that identify the boundaries of the cross-hatched areas. The top boundary (not solid) was generated by interpolation using moment ratio curves for the cases of the form $a = c$, $1 \leq b \leq 25$ and $b = c, 1 \leq a \leq 25$ with $c \in \{0.5, 2, 5, 10\}$. These curves are *not* included in Fig 3.7 to amplify identification of the hatched areas.

A comparison of Figs. 3.6 and 3.7 is quite illuminating. Firstly, Fig. 3.7 indicates that, in terms of moment ratio coverage, the beta family is richer than the STSP family when restricted to the J-shaped forms. The only J-shaped STSP distributions, i.e. the power function distribution and its reflection, are represented within the beta family as indicated by the only common solid lines in Figs. 3.6 and 3.7. The intersection of these lines determines the only other common member of the STSP and beta families, i.e. the uniform|0, 1| distribution. Next, the coverage areas associated with U-shaped forms in Figs. 3.6 and 3.7 are comparable in size, indicating essentially the same flexibility of the STSP and beta families when restricted

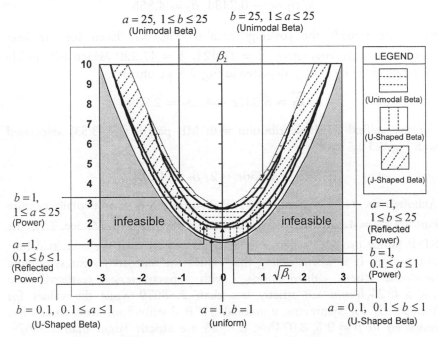

Fig. 3.7 ($\sqrt{\beta_1}, \beta_2$) Moment ratio diagram for beta distributions
with parameters in the range $0.1 \leq a \leq 25, 0.1 \leq b \leq 25$.

to these forms. Finally, possibly most importantly, the coverage area of the
beta family restricted to unimodal forms in Fig. 3.7 is completely contained
within the coverage area of the STSP family restricted to these forms. This
indicates by far a greater flexibility of the STSP family than that of the beta
family when modeling unimodal phenomena for which the mode occurs
not at a support boundary and smooth behavior of the density function at
its mode is not required.

The STSP distribution may thus be considered as an alternative to the
beta distribution especially when sample estimates for skewness $\sqrt{\beta_1}$ and
kurtosis β_2 fall outside the coverage area indicated in Fig. 3.7. For example,
from the data in Fig. 3.1C we estimate

$$\sqrt{\widehat{\beta_1}} = -0.2434, \widehat{\beta_2} = 4.855$$

which are outside the coverage area in Fig. 3.7. Even for the beta distribution with parameters $\widehat{a} = 47.424$, $\widehat{b} = 47.229$ (which fall outside the range given by (3.66)) depicted in Fig. 3.3 we obtain

$$\sqrt{\beta_1} = 8.337e - 4, \beta_2 = 2.939.$$

For the shifted STSP distribution with ML parameters (3.53) associated with Fig. 3.3 we have

$$\sqrt{\beta_1} = 1.836e - 2, \beta_2 = 4.613.$$

Although the estimated shifted STSP distribution does not totally capture kurtosis $\widehat{\beta_2} = 4.855$ observed in the data in Fig. 3.1C (p. 66, Sec. 3.1), the STSP family provides a better fit in terms of $\widehat{\beta_2}$ than the estimated beta distribution. It is worth noting that values for kurtosis β_2 for symmetric unimodal beta distributions with parameters restricted to $1 < a < 25, b = a$ are strictly less than 2.88679 while the values for kurtosis β_2 for symmetric unimodal STSP distributions with parameters restricted to $\theta = 0.5, 3.0745 < n < 25$ are strictly larger than 2.88679. For comparison note that the kurtosis of the normal distribution is 3.

Finally, from Fig. 3.6 it may be observed that the parameter θ of an STSP density *primarily* affects skewness $\sqrt{\beta_1}$ reemphasizing the role of θ as a location parameter. Parameters a and b of the beta distribution affect *both* skewness $\sqrt{\beta_1}$ and kurtosis β_2 in a similar manner and henceforth do not allow for such an interpretation. The reader is encouraged to construct additional comparative diagrams.

3.6 Musings on STSP and Beta Families and Concluding Remarks

In this chapter we have proposed a new family of distributions that possesses certain attractive properties, especially those related to the meaning of its parameters, the structure of its expected value as a function of parameters, a closed form for its cdf, a ML procedure involving only elementary functions and a transparent form of its entropy function. These properties are not fully shared by the beta family of distributions. Similar to the beta family, the new family allows for U-shaped, J-shaped and unimodal forms. The density of the new family is non-smooth (non-differentiable) at θ (see, Eq. (3.11)). It is worth noting that to the best of our knowledge for over 60 years after Karl Pearson's death in 1936 or during even his lifetime no serious attempts have been made to provide an alternative to the beta density of equivalent flexibility. This shows the prominence of Karl Pearson's discovery and a detailed investigation of the beta distribution early in the 20-th century.

For parameter values in the range $1 \leq n \leq 3$ (Eq. (3.11)) the new family displays unimodal forms with a modest peak at θ and provides an attractive alternative to the beta family when smooth behavior at the mode is not a crucial factor while any of the other above mentioned properties are desirable. For parameter values of $n > 3$ the family adds to the existing modeling capabilities of unimodal phenomena on a bounded domain, in particular when peaked data is observed (see, e.g., Fig. 3.3). For parameters values in the range of $0 \leq n < 1$, the family primarily exhibits U-shaped forms similarly to the beta distribution. The only J-shaped distribution within the new family is the power function distribution ($\theta = 1$) (see, Eq. (3.11)) and its reflection ($\theta = 0$) which are also shared by the beta family. The beta distribution thus enjoys greater flexibility among the J-shaped distributions. Consequently, the differences between the proposed family and the beta family of distributions are quite similar to the differences between the Laplace family (which presently is becoming more popular in applications) and the normal family, but both are restricted to a bounded support. (See, e.g., Kotz *et al.* (2001) for a detailed comparison of the Laplace and normal distributions.)

As it was mentioned above in Sec. 3.3 the analysis of the two parameter two-sided power distribution can be extended to the four parameter case including the boundary parameters. Although the ML procedure is somewhat more delicate in this case, it is algorithmically straightforward and painless utilizing modern computational facilities. This will be discussed in Chapter 4. *It is our hope that the introduction of the TSP distribution into statistical theory and practice will contribute to one of the basic goals of applied statistical work: reaching the point when accumulation of data on a specific issue can directly be followed by a sound understanding of the meaning of its parameters.*

Chapter 4

The Two-Sided Power Distribution

In this chapter we present the four-parameter Two Sided Power (TSP) distribution as an alternative to or a substitute for the four-parameter beta distribution in problems involving risk and uncertainty, such as the popular engineering tool "Project Evaluation and Review Technique" (PERT). Properties of the four-parameter TSP distribution follow with minor modifications from the analogous ones of the two-parameter STSP distribution (introduced in Chapter 3) via a linear scale transformation. A maximum likelihood (ML) procedure for the four parameters of the TSP distribution is also derived. It should be mentioned that the ML procedure in this case provides some interesting and unexpected features which were initially observed in a more general setting by Cheng and Amin (1983) and are subject of numerous theoretically oriented publications in statistical and economic journals. Their observations lead to a class of modified ML methods prompted by the behavior of the likelihood function at the endpoints of the domain of variation. Moreover, in problems involving risk and uncertainty, certain parameters of a distribution at hand may be required to be elicited via expert judgment due to lack of appropriate data. Elicitation methods for the parameters of the TSP distribution will be discussed here in the context of a PERT example. The chapter is somewhat long (over 45 pages) and special effort was made to develop its presentation in the most intelligible manner (without sacrificing rigor) by alternating between application (specifically, the PERT example) and theory. Those readers who are more interested in applications may want to just browse through theoretically developments during a first read.

4.1 Introduction: The Four-Parameter TSP Distribution

In recent years two papers dealing with triangular distributions and its extensions have appeared in *The Statistician* (published nowadays by the

Royal Statistical Society, London). The papers by D. Johnson (1997) and N.L. Johnson and Kotz (1999) deal with not very prominent applications of this distribution as an alternative to the four-parameter beta distribution with pdf

$$f_T(t|a, b; \alpha, \beta) = \frac{\Gamma(\alpha+\beta)}{\Gamma(\alpha)\Gamma(\beta)} \frac{(t-a)^{\alpha-1}(b-t)^{\beta-1}}{(b-a)^{\alpha+\beta-1}}$$

$$a \le t \le b,\ \alpha > 0,\ \beta > 0. \tag{4.1}$$

(which involves difficulties related to maximum likelihood estimation of its parameters and has the drawback that the parameters α and β do not possess a clear-cut meaning). D. Johnson's (1997) paper seems to be especially relevant to this chapter. This author specifically suggests the triangular distributions as a proxy to the beta distribution in problems of assessment of risk and uncertainty, such as the Project Evaluation and Review Technique (PERT). Parameters of a triangular distribution have a one-to-one correspondence with an optimistic estimate a, most likely estimate m and pessimistic estimate b of a characteristic under consideration, which provides an intuitive appeal of the triangular distribution (see, e.g., Williams (1992)). Similarly to the beta distribution, the triangular distribution can be positively or negatively skewed (or symmetrical) but always remains unimodal. D. Johnson (1997) points out that there is no triangular distribution which would "reasonably" approximate uniform, J-shaped or U-shaped distributions.

We shall commence with an extension of the three-parameter triangular distribution (to be called the Two-Sided Power (TSP) distribution and denoted $TPS(a, m, b, n)$) with the pdf

$$f_X(x|a, m, b, n) = \begin{cases} \frac{n}{b-a}\left(\frac{x-a}{m-a}\right)^{n-1}, & \text{for } a < x \le m \\ \frac{n}{b-a}\left(\frac{b-x}{b-m}\right)^{n-1}, & \text{for } m \le x < b \\ 0, & \text{elsewhere} \end{cases} \tag{4.2}$$

$$a \le m \le b,\ n > 0,$$

which would seem to be a meaningful alternative to the four-parameter beta distribution (4.1). The meaning of the parameters is as follows: a and b are the endpoints of the support, n is the shape parameter and m is the threshold parameter for a change in the form of the pdf. As in the case of a triangular distribution, parameters a and b may be related to a pessimistic and optimistic estimates of the associated $TSP(a, m, b, n)$ variable. For $n > 1$ $(0 < n < 1)$, m coincides with the mode (anti-mode) of a TSP variable. Since the TSP distribution extends the triangular distribution it inherits its intuitive appeal and naturally interpretable parameters. The pdf (4.2) of X may be derived from that of a random variable Y which is standard two-sided power (STSP) distributed (see Eq. (3.11) in Chapter 3) via the linear scale transformation

$$X = (b - a)Y + a, \quad b > a.$$

Hence, structural forms of the pdf of a TSP distribution are identical to those of the STSP distribution displayed in Chapter 3 in Fig. 3.4 and do allow for J-shaped and U-shaped forms.

The cdf of a $TSP(a, m, b, n)$ distribution follows from (4.2) to be

$$F_X(x|a, m, b, n) = \begin{cases} \frac{m-a}{b-a}\left(\frac{x-a}{m-a}\right)^n, & \text{for } a \leq x \leq m \\ 1 - \frac{b-m}{b-a}\left(\frac{b-x}{b-m}\right)^n, & \text{for } m \leq x \leq b. \end{cases} \quad (4.3)$$

The inverse of the cdf (4.3) can be derived in closed form, yielding

$$F_X^{-1}(y|a, m, b, n) = \quad (4.4)$$
$$\begin{cases} a + \sqrt[n]{y(m-a)^{n-1}(b-a)}, & \text{for } 0 \leq y \leq q \\ b - \sqrt[n]{(1-y)(b-m)^{n-1}(b-a)}, & \text{for } q \leq y \leq 1. \end{cases}$$

where

$$q \equiv Pr(X \leq m) = \frac{m-a}{b-a}, \quad (4.5)$$

regardless of the value of the shape parameter n. Note that the quantity q above may be interpreted as the *relative distance of the mode m to the lower bound a over the support* $[a, b]$ and is reminiscent of the hinge property of STSP distributions (Property 2 in Sec. 3.2.3). Irrespectively of the value of n, the parameter m identifies the q-th percentile of a TSP distribution, where q is given by (4.5). The inverse cdf (4.4) allows straightforward and efficient sampling from the four-parameter TSP distribution in Monte Carlo type uncertainty/risk analyses utilizing a uniform[0, 1] pseudo-random number generator.

The expressions for the mean and the variance can directly be obtained from (4.2) and simplify to

$$E[X] = \frac{a + (n-1)m + b}{n+1} \tag{4.6}$$

and

$$Var(X) = (b-a)^2 \cdot \left\{ \frac{n - 2(n-1)\frac{m-a}{b-a}\frac{b-m}{b-a}}{(n+2)(n+1)^2} \right\}. \tag{4.7}$$

When $n = 1$ we are having a uniform distribution with support $[a, b]$.

4.2 Four-Parameter Maximum Likelihood Estimation

Similar to Sec. 1.5 dealing with ML estimation of the three-parameter triangular distribution, this section involves some non-standard derivations of the ML procedure of the four-parameter TSP distributions which are — as it was already mentioned — closely related to a non-regular case of ML estimation for continuous distributions (see, e.g., Cheng and Amin (1983)).

Let X be a random variable with pdf (4.2). For a random sample $\underline{X} = (X_1, \ldots, X_s)$ with size s from a $TSP(a, m, b, n)$ distribution let the order statistics be $X_{(1)} < X_{(2)} \ldots < X_{(s)}$. Utilizing (4.2), the likelihood for $\underline{X}$ is by definition

$$L(\underline{X}; a, m, b, n) = \tag{4.8}$$

$$\left\{ \frac{n}{b-a} \right\}^s \left\{ \frac{\prod\limits_{i=1}^{r}(X_{(i)} - a) \prod\limits_{i=r+1}^{s} (b - X_{(i)})}{(m-a)^r (b-m)^{s-r}} \right\}^{n-1},$$

where r is implicitly defined by $X_{(r)} \leq m < X_{(r+1)}$, $X_{(0)} \equiv a$ and $X_{(s+1)} \equiv b$. Hence, analogously to the derivation involving the likelihood (1.44) associated with the three-parameter triangular distribution, it follows that for fixed values of a and b, satisfying

$$a < X_{(1)} \text{ and } b > X_{(s)}, \tag{4.9}$$

we have

$$\underset{a \le m \le b, n > 1}{max} \quad L(\underline{X}; a, m, b, n) = \qquad (4.10)$$

$$\underset{n > 1}{max} \quad \left(\frac{n}{b - a}\right)^{s} \left\{ M(a, b, \widehat{r}(a, b)) \right\}^{n-1},$$

where the functions $\widehat{r}(a, b)$ and $M(a, b, r)$ are defined by Eqs. (1.42) and (1.43), respectively, in Chapter 1. (We have elaborated on the behavior of the function $\widehat{r}(a, b)$ in Chapter 1, Sec. 1.5.1.) Analogous to the results in Theorem 3.2 in Chapter 3, the global maximum of (4.8) as a function of m and n only, given fixed values of a and b satisfying (4.9), is attained at

$$\begin{cases} \widehat{m}(a, b) = X_{\widehat{r}(a,b)}; \\ \widehat{n}(a, b) = Max\left\{ -\frac{s}{Log\{M(a,b,\widehat{r}(a,b)\}}, 1 \right\} \end{cases} \qquad (4.11)$$

Hence, as in Sec. 1.5.1 the function $\widehat{r}(a, b)$ is an *index* function indicating at which order statistic the ML estimate of the parameter m is attained as a function of the lower bound a and upper bound b.

As in the preceding chapter we shall now separately consider the two cases of the four-parameter ML estimation procedure, involving the lower bound a and the upper bound b: $0 < n \le 1$ and $n > 1$. The reason is that these two cases yield very different forms of distributions (4.2) and hence different behaviors of the likelihood $L(\underline{X}; a, m, b, n)$ (4.8). Let us first consider the scenario $0 < n \le 1$. For $n = 1$, the pdf (4.2) simplifies to a uniform density with support $[a, b]$ and it is well known that in that case the ML estimates for the lower and upper bounds a and b are:

$$\widehat{a} = X_{(1)} \text{ and } \widehat{b} = X_{(s)} \qquad (4.12)$$

(see, e.g., Devore (2004) or an undergraduate textbook on mathematical statistics). Moreover, setting $a = X_{(1)}$ and $b = X_{(s)}$ into (4.8) with $0 < n < 1$ results in the likelihood $L(\underline{X}; a, m, b, n) \to \infty$ when $a \uparrow X_{(1)}$ (or $b \downarrow X_{(s)}$) for all values of the parameter m such that

$$X_{(1)} < m < X_{(s)}.$$

From here we can conclude that in the case under consideration $(0 < n < 1)$ the ML estimates for a and b are also provided by (4.12). However, the singularity in the likelihood $L(\underline{X}; a, m, b, n)$ (4.8) at

$a = X_{(1)}$ and $b = X_{(s)}$ prevents us to carry out further (stable) ML estimation of the parameters m and n. It should be noted that usually a unimodal *visual* histogram of the data $\underline{X}$ would rule out this situation.

Thus the procedure below for four-parameter ML estimation is intended only for those situations where a value of $n > 1$ could be assessed *a priori*. The unimodal case $(n > 1)$ may be of more practical interest than the case where m is an anti-mode $(0 < n \leq 1)$. (See, for example, the histograms depicting construction data in AbouRizk (1990) and AbouRizk and Halpin (1992).) From (4.10) and (4.11) we have that

$$\underset{S(a,m,b,n)}{max} \left[Log\{L(\underline{X}; a, m, b, n)\} \right] = \qquad (4.13)$$

$$\underset{a < X_{(1)}, b > X_{(s)}}{max} \left[G(a,b) \right],$$

where the set

$$S(a, m, b, n) =$$
$$\{(a, m, b, n)\mid a < X_{(1)}, b > X_{(s)}, a \leq m \leq b, n > 0\}$$

and

$$G(a, b) = s \left\{ Ln \left(\frac{\widehat{n}(a,b)}{b - a} \right) + \left(\widehat{n}(a,b) \right)^{-1} - 1 \right\}. \qquad (4.14)$$

(Compare with Eq. (1.48) in Chapter 1). Note that the function $G(a, b)$ is only defined for values of $a < X_{(1)}$ and $b > X_{(s)}$ (Eq. (4.9)).) Summarizing, the four-dimensional optimization problem of maximizing the likelihood (4.8) reduces to a two-dimensional case of maximizing the function $G(a, b)$ given by (4.14) over (4.9).

As in the case of the function $G(a, b)$ associated with the triangular distribution (Eq. (1.48) in Chapter 1), the function $G(a, b)$ (4.14) involving the TSP distribution is also continuous over its domain (4.9), but the partial derivatives with respect to a or b may not be unique at a finite $(s - 1)$ number of points. The source of non-differentiability at these points originates from the behavior of the index function $\widehat{r}(a, b)$ given by (1.42) as a function of the parameters a and b (See Sec. 1.5.1 for further details). Discarding the points of discontinuity of the function $\widehat{r}(a, b)$, the function

$G(a, b)$ becomes differentiable with respect to a and b. From (4.14) we obtain (recalling the definition of $\widehat{n}(a, b)$):

$$\frac{\partial}{\partial a}G(a, b) = -\frac{\partial M\{a, b, \widehat{r}(a, b)\}/\partial a}{M\{a, b, \widehat{r}(a, b)\}} \times \qquad (4.15)$$
$$\left[\frac{s}{Log[M\{a, b, \widehat{r}(a, b)\}]} + 1\right] + \frac{s}{(b - a)},$$

and

$$\frac{\partial}{\partial b}G(a, b) = -\frac{\partial M\{a, b, \widehat{r}(a, b)\}/\partial b}{M\{a, b, \widehat{r}(a, b)\}} \times \qquad (4.16)$$
$$\left[\frac{s}{Log[M\{a, b, \widehat{r}(a, b)\}]} + 1\right] - \frac{s}{(b - a)},$$

(Compare with Eqs. (1.52) and (1.53) in Chapter 1) where the partial derivatives of $M(a, b, r)$ (Eq. (1.43)) are the same as the ones in case of the three-parameter triangular distribution given by Eqs. (1.54) and (1.55).

The routines $BSearch$ and $ABSearch$ described in Secs. 1.5.2.1 and 1.5.2.2 could be utilized to obtain the ML estimators :

$$\widehat{a}, \ \widehat{b} = \widehat{b}(\widehat{a}), \ \widehat{m}(\widehat{a}, \ \widehat{b}) = X_{(\widehat{r}(\widehat{a}, \ \widehat{b}))} \text{ and} \qquad (4.17)$$
$$\widehat{n}(\widehat{a}, \widehat{b}) = -\frac{s}{Log\{M(\widehat{a}, \widehat{b}, \widehat{r}(\widehat{a}, \widehat{b})\}}$$

by replacing procedures associated with the three-parameter triangular distribution for evaluation of

$$G(a, b) \ (1.48) \ , \ \frac{\partial}{\partial a}G(a, b) \ (1.52) \text{ and } \frac{\partial}{\partial b}G(a, b)(1.53)$$

with procedures to evaluate the function

$$G(a, b) \ (4.14), \ \frac{\partial}{\partial a}G(a, b) \ (4.15) \text{ and } \frac{\partial}{\partial b}G(a, b)(4.16)$$

associated with the four-parameter TSP distribution and adding the following step to the routine $ABSearch$ described in Sec. 1.5.2.2 after Step 7 :

$$Step\,8: \ n_k = \frac{-s}{Log(M_k)}.$$

4.2.1 An illustrative example

Figure 4.1 provides the form of the function $G(a, b)$ for a civil engineering data set in Table 1.2 in Chapter 1 (source: AbouRizk (1990)).

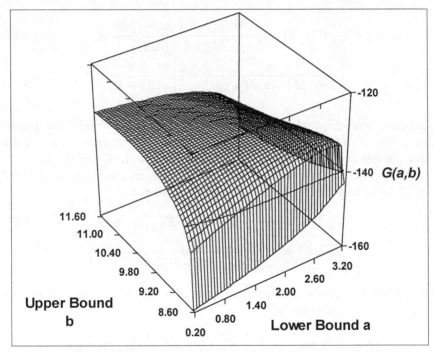

Fig. 4.1 The function $G(a, b)$ given by (4.14) for the data in Table 1.2 in Chapter 1.

For our construction engineering data presented in Table 1.2 the modified routine $ABSearch$ results in the following ML estimates :

$$\widehat{a} = 1.632; \widehat{b} = 9.573;$$
$$\widehat{m}(\widehat{a}, \widehat{b}) = 5.9; \widehat{n}(\widehat{a}, \widehat{b}) = 3.463.$$

(4.18)

Figure 4.2A (Figure 4.2B) depicts a likelihood profile of the function $G(a, b)$ displayed in Fig. 4.1 for the data in Table 1.2 as function of the parameter a (parameter b) for different (fixed) values of the parameter b (parameter a). (Compare with Fig. 1.8 in Chapter 1.) Note the behavior of $G(a, b)$ for $b = 8.6$ (for $a = 3.2$) in Fig. 4.2A (Fig. 4.2B). The ML estimates $\widehat{a} = 1.632$ and $\widehat{b} = 9.573$ are indicated by means of a vertical dotted line in Figs. 4.2A and 4.2B, respectively.

From the ML estimates in (4.18) and the data in Table 1.2 it follows that

$$\widehat{a} = 1.632 < X_{(1)} = 3.2 \text{ and } \widehat{b} = 9.573 > X_{(s)} = 8.6.$$

Compare these values with those in Eq. (1.50) in Chapter 1 and observe that the lower bound (upper bound) ML estimate involved with the triangular distribution is larger (smaller) than the corresponding TSP one in (4.18). Figure 4.3A depicts the empirical pdf of the data in Table 1.2 together with the four-parameter ML fitted TSP distribution with ML parameter values specified in (4.18) (Compare with Fig. 1.6 in Chapter 1). Figure 4.3B also displays a four-parameter beta distribution (4.1) fitted to the data in Table 1.2 via a least squares method proposed in AbouRizk (1990) by minimizing

$$\sum_{i=1}^{s} \{F(X_{(i)}|a, b; \alpha, \beta) - \frac{i}{n+1}\}^2,$$

(4.19)

where $F(X_{(i)}|a, b; \alpha, \beta)$ is the cdf associated with the pdf (4.1) (also known as the incomplete beta function). Its fitted parameters are

$$\widehat{a} = -9.097; \widehat{b} = 16.588; \widehat{\alpha} = 75.076; \widehat{\beta} = 55.530.$$

(4.20)

Note the much wider support $[\widehat{a}, \widehat{b}]$ for the fitted beta distribution in (4.20) than that of the ML fitted TSP distribution in (4.18) (although visually it is not easy to observe by comparing Figs. 4.3A and B). We utilize the least squares method above to fit the beta distribution since it allows straightforward implementation in the software tool Microsoft EXCEL

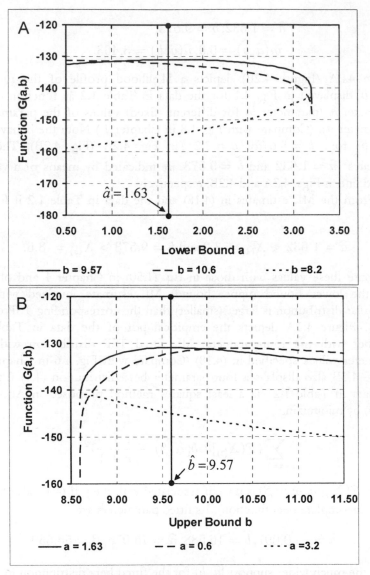

Fig. 4.2 Profiles of the function $G(a, b)$ given by (4.14) for the data in Table 1.2:
Graph A: as a function of the lower bound a; Graph B: as a function of the upper bound b.
The ML estimates $\widehat{a} = 1.632; \widehat{b} = 9.573$ in (4.18) are indicated by means
of a vertical dotted line in Figs. 4.2A and 4.2B, respectively.

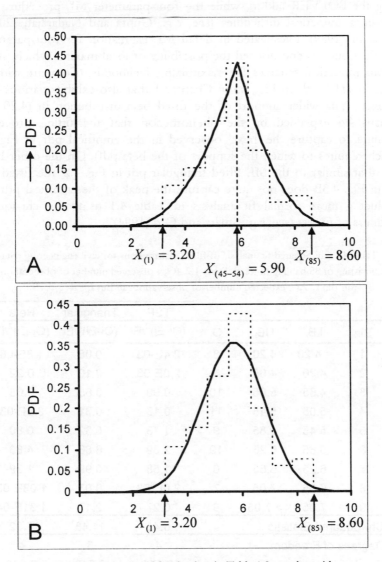

Fig. 4.3 Empirical pdf for the data in Table 1.2 together with
A: an ML fitted four-parameter TSP distribution with parameters $\widehat{a} = 1.632$,
$\widehat{m} = X_{(45-54)} = 5.9$, $\widehat{b} = 9.573$ and B: a least squares fitted beta distribution with
parameters $\widehat{a} = -9.097$; $\widehat{b} = 16.588$; $\widehat{\alpha} = 75.076$; $\widehat{\beta} = 55.530$.

using the SOLVER add-in, while the four-parameter ML procedure may encounter numerical difficulties (see, e.g., Gupta and Nadarajah (2004)). Carnahan (1989) investigated in detail ML estimation for four-parameter beta distributions and noticed the possibility of local maxima which plague various numerical schemes for maximizing likelihoods. (Compare with the function $H(X ; \theta)$ in Fig. 1.5 in Chapter 1 that also exhibits various local maxima.) The wider support of the fitted beta distribution in (4.20) can perhaps be explained by the phenomenon that the fitting procedure attempts to capture the 'peak' observed in the empirical pdf in Fig. 4.3 (which requires to widen the support of the beta pdf). On the other hand, note that similar to the ML fitted triangular pdf in Fig. 1.2, the fitted beta pdf in Fig. 4.3B does not quite capture the peak of empirical pdf. We conduct a more formal fit analysis in Table 4.1 using the chi-squared goodness of fit test (see, e.g., Stuart and Ord (1994)).

Table 4.1 Chi-Squared fit analysis of fitted distributions for civil engineering data consisting of 85 hauling times in Table 1.2. (O_i : observed number of observations in Bin i, E_i : expected number of observations in Bin i, $i = 1, \ldots 9$.

Bin	LB_i	UB_i	O_i	TSP $(Oi-Ei)^2/Ei$	Triangular $(Oi-Ei)^2/Ei$	Beta $(Oi-Ei)^2/Ei$
1	< 4.20	4.20	8	2.4E-03	0.05	1.8E-06
2	4.20	4.65	6	2.0E-03	0.10	0.22
3	4.65	5.05	10	0.90	0.65	0.08
4	5.05	5.45	11	0.13	0.31	3.9E-03
5	5.45	5.85	9	1.13	0.32	0.80
6	5.85	6.25	19	2.39	6.86	4.83
7	6.25	6.65	6	1.58	0.99	1.39
8	6.65	7.05	7	2.0E-03	0.03	1.03E-03
9	7.05	> 7.05	9	0.27	2.17	1.31E-04
Chi-Squared Statistic				6.41	11.48	7.32
Degrees of Freedom				4	5	4
P-value				0.171	0.043	0.120

In Table 4.1 the chi-squared statistic

$$\sum_{i=1}^{9} \frac{(O_i - E_i)^2}{E_i}$$

is calculated utilizing the 9 bins. (Note that, $9 \in [\sqrt{85}, 85/5]$ as suggested by Banks *et al.* (2001).) With exception of the first and last bin, each bin has width 0.45. The number of observations O_i for the data in Table 1.2 is provided in Table 4.1, at least 6 in each bin. The corresponding values $E_i, i = 1, \ldots, 9,$ for the expected number of observations in Bin i are obtained using the formula

$$E_i = 85\{F(UB_i|\widehat{\Theta}) - F(LB_i|\widehat{\Theta})\},$$

where $F(\cdot |\widehat{\Theta})$ is the theoretical cdf, $\widehat{\Theta}$ are the corresponding parameter estimates for each fitted distribution given by (4.18), (1.50) and (4.20) for the TSP, triangular and beta distributions respectively, and the bin boundaries $(LB_i, UB_i]$ are provided in Table 4.1. Similarly to the fit analysis conducted in Table 3.2 in Chapter 3 (involving financial data) we observe here that Bin 6 containing the peak of the empirical pdf in Figs. 1.6 and 4.2 contributes the most to the chi-squared statistic values in Table 4.1. From the p-values in Table 4.1 we conclude that while the beta distribution with parameters (4.20) performs better than the triangular distribution with parameters (1.50) for the data in Table 1.2, it follows that the TSP distribution with parameters (4.18) is, in turn, a superior fit than the beta distribution.

In civil engineering applications it is perhaps not uncommon to use estimation procedures to determine a distribution's parameters (see, e.g. AbouRizk and Halpin (1992)). Moreover expert judgment elicitation is also used to estimate a distribution's parameter in, for example, a PERT context. (The PERT technique is used, amongst others, to determine the uncertainty in the completion time of large engineering projects.)

4.3 Elicitation Methods for TSP Distributions

The phenomenal advances in quantitative methodology and their rapid penetration into applied sciences and engineering during the last few decades have resulted in a re-assessment of the scope and the nature of an expert's activities. As a rule experts are classified into two, usually unrelated, groups: *Substantive* experts (also referred to as technical experts or domain experts) who are knowledgeable of the various aspects of the subject matter under consideration and *normative* experts possessing adequate knowledge about the appropriate quantitative analysis techniques (see, e.g., Pulkkinen and Simola (2000) for more details). In the absence of data, expert judgment concerning the quantities of interest are elicited from substantive experts and equated by normative experts to their theoretical expressions (see, e.g., Cooke (1991)). In the context of the PERT analysis, for example, substantive experts are used, often by necessity, to elicit the so-called optimistic estimate $\widehat{a}$, most likely estimate $\widehat{m}$ and pessimistic estimate $\widehat{b}$ of an unknown activity duration T in a PERT network (see, e.g., the basic paper by Moder and Rodgers (1968) that appeared some 35 years ago and still provides useful interpretations).

The decision concerning the type of the distribution chosen to model the uncertainty of an activity duration at hand (for which traditionally the beta and triangular distributions have been used) and on the manner to estimate their parameters utilizing $\widehat{a}$, $\widehat{m}$ and $\widehat{b}$, is usually carried out by a normative expert (who may not necessarily be experienced in qualitative and technical details related to the completion of the activity under consideration). Modeling assumptions of this kind may often be inevitable due to lack of data. It is, however, highly desirable that their impact on the analyses to be carried out will be minimal.

Malcolm *et al.* (1959) utilized the estimates $\widehat{a}$, $\widehat{m}$ and $\widehat{b}$ to fit a four-parameter $Beta(a, b; \alpha, \beta)$ distribution given by the pdf (4.1). Next, by equating $a = \widehat{a}$ and $b = \widehat{b}$ and using the method of moments technique by setting

$$\begin{cases} \widehat{E}[T] = \frac{\widehat{a}+4\widehat{m}+\widehat{b}}{6} \\ V\widehat{ar}[T] = \frac{1}{36}(\widehat{b} - \widehat{a})^2, \end{cases} \tag{4.21}$$

and equating them to their population expressions

$$\begin{cases} E[T] = \frac{\alpha}{\alpha+\beta}(b - a) + a \\ Var[T] = (b - a)^2 \frac{\alpha\beta}{(\alpha+\beta+1)(\alpha+\beta)^2} \end{cases} \tag{4.22}$$

Malcolm provides estimates of the parameters α and β in (4.1). The elicitation procedure above is classified as an *indirect* elicitation procedure for the parameters α and β and as a *direct* elicitation procedure for the parameters a and b (see, e.g., Cooke (1991)). This attractive approach allows practitioners dealing with risk and uncertainty to overcome difficulties associated with the interpretation of parameters (in this case α and β). Malcolm *et al.*'s (1959) recommendation to utilize (4.21) and (4.22) for an indirect elicitation of α and β, however, resulted in a heated discussion regarding their applicability and appropriateness that has been ongoing for some 40 years by now (see, e.g., Clark (1962), Grubbs (1962), Moder and Rodgers (1968), Keefer and Verdini (1993), Kamburowski (1997), Lau *et al.* (1998), among others). Kamburowski (1997) emphasizes that:

"Since PERT development, there has been controversy associated with the validity of" (4.21) ... "Both PERT estimates have been roundly criticized, large estimation error have been attributed to them, and many refinements have been proposed to replace them. Despite the criticisms and the abundance of new estimates, the PERT mean and variance can be found in almost every textbook on OR/MS and P/OM, and are employed in much project management software. Unfortunately, these formulas are mostly presented without any additional assumptions which leads to so many misunderstandings" ... *"Undoubtedly, all the above have diminished the practical interest in the original PERT".*

Kamburowski (1997) continues by presenting new theoretical validations for the use of (4.21) based on a kurtosis argument. (See, the same paper for a comprehensive bibliography dealing with this PERT "controversy".)

Perhaps to resolve this "controversy", D. Johnson (1997) specifically suggested that the three parameter triangular distribution with the pdf (1.6) given in Chapter 1 be used in PERT analyses as an alternative to the beta distribution (and perhaps others before him as well), since the parameters a, m and b in (1.6) are a one-to-one correspondence with the estimates $\widehat{a}, \widehat{m}$ and $\widehat{b}$ (that have been provided by the substantive expert). Hence, all the parameters of a triangular distribution (due to their physical interpretation) may be elicited *directly* from a substantive expert. This gives to the triangular distribution substantial intuitive appeal as far practical applications are concerned (Williams (1992)).

Extending Johnson's approach, Van Dorp and Kotz (2004, 2002b) suggested the use of the four-parameter $TSP(a, m, b, n)$ distribution (which generalizes the triangular distribution (1.6)) with the pdf (4.2) as a proxy for the beta distribution, in particular in problems involving an assessment of risk and uncertainty (such as PERT). The TSP density (4.2) coincides for $n = 2$ with the triangular density (1.6). Expressions for the population mean and the variance for the TSP pdf (4.2) are given by Eqs. (4.6) and (4.7), respectively. (Note that in (4.2), (4.6) and (4.7) we use the notation X to indicate a random variable with a TSP distribution — or, for $n = 2$ the triangular distribution — whereas the random variable T in (4.1), (4.21) and (4.22) indicates a random variable obeying a beta distribution. Both the letters X and T have the same physical meaning designating models for uncertainty in an activity duration.)

4.3.1 Indirect elicitation of the shape parameter n

The use of the $TSP(a, m, b, n)$ pdf instead of the triangular one in the above context, requires the elicitation of an estimate $\widehat{n}$ of the shape parameter n from the substantive expert (in addition to the already directly elicited estimates $\widehat{a}$, $\widehat{m}$ and $\widehat{b}$). From the expression for the mean (4.6) it follows that $(n + 1)$ may be interpreted as the sample size of a virtual sample with $(n - 1)$ observations m, with two additional observations a and b. Hence, one may indirectly elicit $\widehat{n}$ by asking an expert for the *relative importance* of the elicited most likely value $\widehat{m}$ as compared to the bounds $\widehat{a}$ or $\widehat{b}$, where a value of 1 indicates the same importance and values greater (less) than 1 indicate larger (lesser) importance. Suppose an expert assigns the relative importance of the most likely value m to be $y = (n - 1)$, it then follows from the interpretation above that

$$n = y + 1. \tag{4.23}$$

Hence, if an expert responds that the most likely estimate $\widehat{m}$ is equally important compared to the bound estimates $\widehat{a}$ or $\widehat{b}$ (i.e. $\widehat{y} = 1$), this implies from (4.23) that a triangular distribution ($\widehat{n} = 2$) appropriately models the expert's uncertainty. If an expert responds that the most likely estimate m is more (correspondingly, less) important than the bounds a or b, the elicitation will yield a TSP distribution with variance smaller (correspondingly, larger) than that of the triangular distribution. An expert would have to assign relative importance $y = 4$ for the mean of the resulting TSP variable $E[X]$ to agree with $E[T]$ in (4.21) proposed by Malcolm *et al.* (1959). Finally, a relative importance of *zero* of $\widehat{m}$ (compared to the bound estimates $\widehat{a}$ or $\widehat{b}$) indicating no confidence whatsoever in the most likely estimate $\widehat{m}$, results via (4.23) in $\widehat{n} = 1$ and the TSP pdf (4.2) reduces to a uniform pdf on the interval $[\widehat{a}, \widehat{b}]$.

While the elicitation approach above utilizes a comparison approach (similar to the popular paired comparison elicitation techniques in psychological scaling models, see, e.g., Cooke (1991)), the concept of relative importance to elicit the shape parameter n may be considered somewhat difficult to interpret by a substantive expert. In Sec. 4.3.3 an elicitation procedure will be developed for the shape parameter n (and the lower bounds a and b) that utilizes quantile estimates from the substantive expert. An advantage of eliciting quantile estimates, instead of *relative importance*, is that they allows for the use of betting strategies in an indirect elicitation procedure (see, e.g., Cooke (1991)).

4.3.2 The PERT "controversy" via an illustrative example

Before continuing with the development of an indirect elicitation procedure involving the bound parameters a, b in addition to the shape parameter n, it may be useful to illustrate via an example the reasons that the simultaneous use of (4.21) and (4.22) have been "controversial" and non-rigorous from a statistical point of view.

It follows from (4.2) that for a triangular distribution $(n = 2)$ the population mean $E[X]$ may over- or under estimate the estimator for $E[T]$ as given in (4.21) depending on whether or not the threshold parameter m is less or greater than the midpoint $(a + b)/2$. Note, however, that for a TSP distribution with $n = 5$, the estimated mean value $E[T]$ given in (4.21) and the population mean value $E[X]$ given in (4.6) actually coincide. On the other hand, it follows from (4.21) and (4.7) that in the case of a triangular distribution $(n = 2)$

$$Var[T] = \frac{1}{36}(b-a)^2 < \frac{3}{72}(b-a)^2 \leq Var[X] \leq \frac{1}{18}(b-a)^2 \quad (4.24)$$

and for a TSP distribution with $n = 5$ we have correspondingly

$$\frac{1}{84}(b-a)^2 \leq Var(X) \leq \frac{5}{252}(b-a)^2 < Var[T] = \frac{1}{36}(b-a)^2. \quad (4.25)$$

Hence, from Eq. (4.24) (Eq. (4.25)) it follows that $Var[X]$ of a triangular distribution (a TSP distribution with $n = 5$) is always larger (less) than the estimator for $Var[T] = (\widehat{b} - \widehat{a})^2/36$ in (4.21), regardless of the values of the parameters a, m and b (see also Van Dorp and Kotz (2002b), (2004a)).

We shall demonstrate the effect of the previous conclusions on a small 18-activity project network $\mathcal{P}$ depicted in Fig. 4.4 in the ship building domain described in Taggart (1980). The parameters a, m, b of triangular distributions (1.6) for the activities in the project network $\mathcal{P}$ are provided in Table 4.2, which also contains a column for the parameter values $n = 5$ in case $TSP(a, m, b, 5)$ distributions are used to model the activity duration uncertainty. In addition, Table 4.2 depicts columns for the estimated mean $E[T]$ and variance $Var[T]$ that follow from (4.21) and the values for a, m and b. Finally, Table 4.2 contains columns for the values of the parameters α and β of beta distributions with support $[a, b]$ that were solved from the values of $E[T]$ and $Var[T]$ in Table 4.2 using the method of moments.

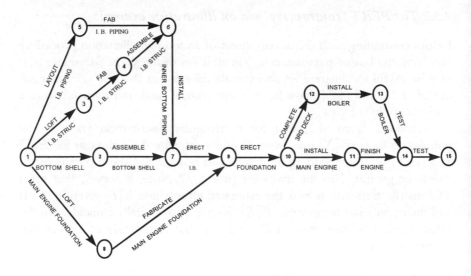

Fig. 4.4 Example project network $\mathcal{P}$ for production process.

Next, we successively sample (25000 times) from the uncertainty distributions of the activity durations (triangular, beta or TSP(5)) and each time subsequently evaluate the completion time of project the project network $\mathcal{P}$ using the Critical Path Method (CPM) (see, e.g. Winston (1993)). This technique is known as the Monte Carlo approach (see, e.g. Vose (1996)) and allows us to calculate the mean, standard deviation, minimum and maximum of the completion time distribution up to a desired accuracy level. Table 4.3 contains the resulting values for these statistics utilizing the three different approaches described above for modeling activity uncertainty.

Table 4.2 Parameters for modeling the uncertainty in activity durations for the project network in Figure 1 A: via (4.2) with $n = 5$ (Triangular), B: via (4.1) − utilizing (4.21) and (4.22) using the method of moments technique − (Beta), C: via (4.2) with $n = 5$, (TSP(5)). Recall that for $n = 5$, the estimated mean value $E[T]$ given in (4.21) and the population mean value $E[X]$ given in (4.6) actually coincide.

ID	Activity Name	a	m	b	n	E[T]	Var(T)	α	β
1	Shell: Loft	22	25	31	5	25.5	2.3	2.94	4.62
2	Shell: Assemble	35	38	43	5	38.3	1.8	3.23	4.52
3	I.B.Piping: Layout	19	22	47	5	25.7	21.8	1.32	4.21
4	I.B.Piping: Fab.	6	7	15	5	8.2	2.3	1.34	4.24
5	I.B.Structure: Layout	23	24	30	5	24.8	1.4	1.56	4.40
6	I.B.Structure: Fab.	14	18	24	5	18.3	2.8	3.40	4.44
7	I.B.Structure: Assemb.	9	14	20	5	14.2	3.4	3.74	4.22
8	I.B.Structure: Install	5	7	13	5	7.7	1.8	2.33	4.67
9	Mach Fdn. Loft	26	28	33	5	28.5	1.4	2.59	4.67
10	Mach Fdn. Fabricate	20	35	51	5	35.2	26.7	3.91	4.08
11	Erect I.B.	27	30	37	5	30.7	2.8	2.70	4.66
12	Erect Foundation	5	7	14	5	7.8	2.3	2.13	4.64
13	Complete #rd DK	4	5	9	5	5.5	0.7	1.97	4.59
14	Boiler:Install	6	7	11	5	7.5	0.7	1.97	4.59
15	Boiler:Test	9	10	15	5	10.7	1.0	1.73	4.49
16	Engine: Install	6	7	15	5	8.2	2.3	1.34	4.24
17	Engine: Finish	19	20	26	5	20.8	1.4	1.56	4.40
18	FINAL Test	13	15	24	5	16.2	3.4	1.84	4.54

Note that the results in Table 4.3 regarding the standard deviation are consistent with the conclusions (4.24) and (4.25) related to $Var[T]$ and $Var[X]$. In addition, from the values for a, m and b in Table 4.2 it follows that for all 18 activities the mode m is less than the midpoint $(a + b)/2$. Hence, $E[T]$ in (4.21) for all 18 activities underestimates the mean value $E[X]$ of the triangular case (1.6) in Chapter 1 resulting in a noticeably smaller value of the mean project completion time for the beta case (150.5) in the second row of Table 4.3 than for the triangular case (157.3). The mean values $E[T]$ for the activities in Table 4.2 and that for $E[X]$ of a $TSP(a, m, b, 5)$ variable coincide, resulting in a close value for the mean project completion time in the second and third rows of Table 4.3 (150.5 and 150).

Table 4.3 Mean and standard deviation of the project completion time
distribution using triangular (suggested by Johnson (1997)), beta distributions
(suggested by Malcolm (1959)) and TSP (5), (suggested by Van Dorp and Kotz (2002b)).

	Mean	Standard Deviation	Min	Max
Triangular	157.3	6.0	136.9	181.1
Beta	150.5	4.7	134.4	169.4
TSP(5)	150.0	3.5	139.4	164.4

From the analysis in Table 4.3 we may conclude that the choice of a normative expert to estimate the parameters of a beta distribution via (4.21) and (4.22), or utilize a triangular or TSP distributions using solely the "raw" estimates $\widehat{a}$, $\widehat{m}$ and $\widehat{b}$ affects the PERT analysis to a level that he/she may not be quite comfortable with. Consequently, the suggestions of D. Johnson (1997) and that of Van Dorp and Kotz (2002b) seem only to augment the existing inconsistencies related to the setup given by (4.21), as proposed by Malcolm *et al.* (1959). Hence, it would be natural to inquire what additional information is required from a substantive expert regarding the activity durations, so that the modeling choices by a normative expert of uncertainty distribution of activity durations will have a lesser or preferably minuscule effect on the PERT analysis. We shall attempt to provide an answer to this question in Sec. 4.3.3.

4.3.2.1 Sources for the PERT "controversy" in Sec. 4.3.2

While the example in the previous section demonstrates the reason for a PERT "controversy" to exist surrounding the use of (4.21) and (4.22), it does not fully identify its sources. It may be useful for clearer understanding to dwell briefly on the objections to using estimators (4.21) which endured vitality for over 40 years, since it is a rather important caveat in applying the standard PERT analyses. We shall concentrate on the illustrative example where a substantive expert has assessed the values of the unknown parameters a, m and b to be $\widehat{a} = 6$, $\widehat{m} = 7$ and $\widehat{b} = 15$.

A principle source for the "controversy" surrounding (4.21) stems most probably from the fact that there are indeed infinitely many unimodal pdf's with support $[\widehat{a}, \widehat{b}]$ having a mode at $\widehat{m}$. Evidently, these pdf's possess various different uncertainty characteristics. Figure 4.5 (Figure 4.6) depicts several members within the beta (TSP) family all satisfying the $\widehat{a} = 6$, $\widehat{m} = 7$ and $\widehat{b} = 15$ assessments. An attentive reader would notice that the solid curve in Fig. 4.5 (Example A) has the mode m slightly less than $\widehat{m} = 7$. In fact, Example A in Fig. 4.5 presents the beta distribution which is obtained when (4.21) and (4.22) are used to solve for the parameters α and β, setting $a = 6$ and $b = 15$. The set-up (4.21) when used to solve for the beta parameters results in the value $m = 6.86$ and does not precisely represent the originally assessed value for the mode $\widehat{m} = 7$. (Note that both Figs. 4.5 and 4.6 include the uniform distribution with support $[6, 15]$ − a distribution which appears in both TSP and Beta families − and may thus be viewed a "degenerate" case satisfying $m = 7$.)

Table 4.4 provides the values for skewness $\beta_1 = \mu_3^2/\mu_2^3$ and kurtosis $\beta_2 = \mu_4/\mu_2^2$, $E[X] = \mu_1$, where $\mu_k = E[X - \mu_1]^k$ (for $k > 1$) for the density functions in Figs. 4.4 and 4.5. The values for the kurtosis (skewness) in the examples in Figs. 4.4 and 4.5 range from 1.80 to 4.14 (from 0.08 to 1.30). (The kurtoses of the curves depicted in examples B in Figs. 4.4 and 4.5 are identical, while skewnesses are very close to each other.) From Figs. 4.5 and 4.6 and Table 4.4 it follows that it may not be possible (for a normative expert) to devise a methodology (along the lines of (4.21)) for describing the uncertainty of an activity solely based on the estimates $\widehat{a}, \widehat{m}$ and $\widehat{b}$ (which are provided by a substantive expert) that will also be fully consistent. Indeed, whatever is the motivation of a normative expert to "pick" a particular member of a family of distributions, the choice would certainly affect to a large extent the PERT analysis to be carried out (corresponding to the very substantial differences similar to those amongst the distributions presented in Figs. 4.5 and 4.6 − all of them with the same a, m and b values).

Another possible source for inconsistency stemming from the use of (4.21) is that it is very unlikely that a substantive expert be able to accurately assess the actual lower bound a (i.e. the 0-th percentile) and the upper-bound b (the 100-th percentile). Indeed, these extreme values (rarely occurring in practice) are very likely to fall outside the realm of his/her experience in spite of his/her knowledge of the activity at hand. Therefore early PERT practitioners have taken the liberty (with ample justification) to replace the lower bound estimate $\widehat{a}$ (upper bound estimate $\widehat{b}$) by the p-th

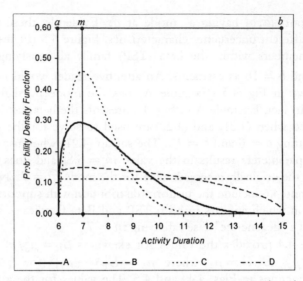

Fig. 4.5 Examples of members of the beta family with support $[6, 15]$:
A: $\alpha = 1.343$, $\beta = 4.237$ $(mode = 6.863)$, B: $\alpha = 1.063$, $\beta = 1.500$ $(mode = 7)$,
C: $\alpha = 2.125, \beta = 10$ $(mode = 7)$, D: $\alpha = 1, \beta = 1$ $(mode = 7)$.

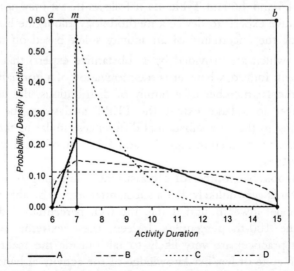

Fig. 4.6 Examples of members of the TSP family with support $[6, 15]$
and mode $m = 7$: A: $n = 2$, B: $n = 1.442$, C: $n = 4$, D: $n = 1$.

Table 4.4 Skewness β_1 and Kurtosis β_2 for the distributions in Figs. 4.5 and 4.4.

	Beta		TSP	
	Skewness	Kurtosis	Skewness	Kurtosis
A	0.674	3.194	0.291	2.400
B	0.082	2.022	0.084	2.022
C	0.768	3.679	1.309	4.145
D	0.000	1.800	0.000	1.800

$((1 - p)$-th) percentile, setting $a_p = \widehat{a}$ $(b_{1-p} = \widehat{b})$. Motivated by statistical hypotheses testing tradition, popular values for p were chosen to be 0.01, 0.05 and 0.10 and sensitivity analyses were conducted on these values (see, e.g., Moder and Rodgers (1968)). However, a *normative* expert, although possessing substantial knowledge of the related quantitative techniques, may very well consider this assessment of p to be outside his/her competence.

Instead, it would seem that eliciting a_p and b_{1-p} from a *substantive* expert (rather than inquiring for the extreme lower and upper bounds) is a more appropriate approach. It has been verified, however, that assessment of percentiles in the vicinity of extremes, such as the 0.01 and 0.99 percentiles (Alpert and Raiffa (1982)) is also (as is the case with the lower and upper bounds) very often beyond our accumulated experience, since the extreme fractiles are associated with very rare events (Keefer and Verdini (1993)). The latter authors observed that the "moderate" 0.10 and 0.90 quantiles have been found to be more reliable than the extreme 0.01 and 0.99 percentiles (Selvidge (1980)) or even the "intermediate" ones 0.05 and 0.95 fractiles (Davidson and Cooper (1980)). Modifications of method of moment type estimates (4.21) have been provided to accommodate the use of the appropriate a_p and b_{1-p} (see, e.g., Moder and Rodgers (1968) and Keefer and Verdini (1993)).

Incorporation of these modifications is, however, still not sufficient to totally quell the existing "controversy". Indeed, Fig. 4.7 below depicts a member of the four-parameter beta family with support $[6, 15]$, that of a three parameter triangular distribution with support $[5.464, 12.452]$, a two-parameter uniform distribution with support $[6, 11]$ and a four-parameter TSP distribution with $n = 7$ and support $[2.687, 20.891]$. All these four distributions in Fig. 4.7 satisfy the constraints set by the lower percentile $a_{0.10} = 6.5$, the most likely value $m = 7$ and the upper percentile

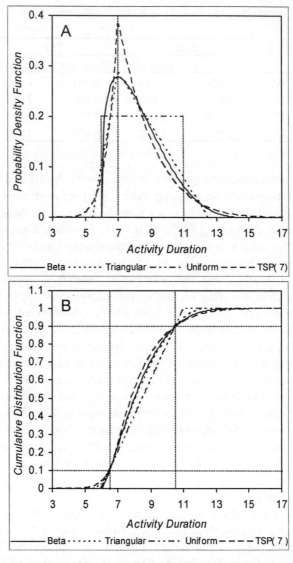

Fig. 4.7 Beta, triangular, uniform and TSP distributions satisfying the constraints set by $a_{0.10} = 6.5$, $m = 7$, and $b_{0.90} = 10.5$. Beta with parameters: $a = 6$, $b = 15$, $\alpha = 1.384$, $\beta = 4.071$; Triangular: $a = 5.464$, $m = 7$, $b = 12.452$; Uniform: $a = 6$, $b = 11$; TSP: $a = 2.687$, $m = 7$, $b = 20.891$, $n = 7$. Graph A: probability density functions (pdf), Graph B: cumulative distribution functions (cdf).

$b_{0.90} = 10.5$, while clearly noticeable differences in their support are evident. Note that the support of the peaked TSP distribution with $n = 7$ is by far the widest, while for the flat uniform distribution it is the narrowest. (In fact it will be shown, in the next section, that the support of a $TSP(a, m, b, n)$ distribution satisfying $a_{0.10} = 6.5$, $m = 7$ and $b_{0.90} = 10.5$ can be made arbitrarily large by letting $n \to \infty$.)

Table 4.5 provides the values for the mean, variance, skewness and kurtosis for these four distributions with the same $a_{0.10} = 6.5$, $b_{0.90} = 10.5$ and most likely value $m = 7$. Note that the very large values of the variance, skewness and kurtosis correspond to the TSP distribution while the smallest values are obtained for the uniform one. On the other hand, the mean values in Table 4.5 behave in a opposite manner. The differences amongst the distributions in Figure 4.7 (and analogously Table 4.5) would certainly affect a PERT analysis. From the above analysis one concludes that eliciting a lower percentile a_p, a most likely estimate m and an upper estimate b_{1-p} is, unfortunately, also not quite sufficient to fully describe an uncertainty distribution of an activity duration in a PERT context.

Table 4.5 Lower Bound, Upper Bound, Mean, Variance,
Skewness and Kurtosis for the distributions in Fig. 4.7.

	Beta	Uniform	Triangular	TSP(7)
Lower Bound	6.00	6.00	5.46	2.69
Upper Bound	15.00	11.00	12.45	20.89
Mean	8.28	8.50	8.31	8.20
Variance	2.38	2.08	2.25	2.78
Skewness	0.60	0.00	0.21	1.25
Kurtosis	3.08	1.80	2.40	4.60

4.3.2.2 Attempts to improve the mean and variance estimation using expert judgment

A number of investigators arrived at a similar conclusion as the one above and suggested the elicitation of three (Pearson and Tukey (1965)), five (Alpert and Raiffa (1982)), seven (Selvidge (1980)) or even nine quantiles (Lau et al. (1998)). Numerous studies have been carried out with the aim to investigate relationships amongst these quantiles to accurately assess the mean and the variance of an activity duration across a wide variety of families of distributions. A most popular approach is the time-honored extended Pearson-Tukey method (developed in the 60's) for estimating the mean μ and the variance σ^2 of a general distribution via the following relationships:

$$\hat{\mu} = 0.185x_{0.05} + 0.630x_{0.50} + 0.185x_{0.95} \tag{4.26}$$

$$\hat{\sigma} = 0.630(x_{0.50} - \hat{\mu})^2 + 0.185\{(x_{0.05} - \hat{\mu})^2 + (x_{0.95} - \hat{\mu})^2\}$$

(in an obvious notation).

The motivation to assess the mean and the variance of the activities in a PERT network as accurately as possible stems from the fact that with an *assumption of independence* amongst the activity durations, these two values allow us to approximate the mean completion time of a project network and its variance by means of a closed form expression (see, e.g., Keefer and Verdini (1993)). However, the independence assumption of activities within a PERT network is usually stipulated primarily for the sake of mathematical convenience and is specious at best (as pointed out by Duffey and Van Dorp (1997) and Van Dorp (2004), among others). To the best of our knowledge for statistically dependent activity durations, no approximating closed form expressions for the mean and the variance of the minimal completion time of a PERT network are available, that utilize solely the mean and the variance of the activity durations. One is thus compelled here to resort here to the eclectic Monte Carlo simulations of the project network by repeated sampling from the distributions of the activities and then successively applying the CPM to each sample (see, e.g., Vose (1996)). The latter requires a complete description of the whole distribution function of activity durations (and not just their mean and variance). Modern computational facilities present no obstacle for conducting simulations of large scale PERT networks without imposing the independence assumption amongst the activities (Van Dorp (2004)). Hence, once again a normative expert is required to decide on a family of distributions to model the uncertainty of an activity duration, be it beta, triangular, TSP or any other appropriate distribution.

We feel that it is useful to remind the reader that an important reason that the triangular distribution enjoys intuitive appeal amongst practitioners and engineers is that its parameters a, m and b have an one-to-one correspondence to optimistic ($\widehat{a}$), most likely estimate ($\widehat{m}$) and pessimistic ($\widehat{b}$) estimates (D. Johnson (1997) and Williams (1992)). In addition, these parameters are of the same dimension as the quantity of interest and the probability mass to the left of the most likely value m turns out to be the relative distance from the mode m to the lower bound a relative to the whole support $[a, b]$, i.e. $(m - a)/(b - a)$. These properties are inherited by the $TSP(a, m, b, n)$ distributions as well, for all values of the shape parameter n. Hence, it seems to be a reasonable approach to build on the intuitive appeal of the triangular distribution and specify what additional information needs to be assessed by the substantive expert (besides the values of a_p, m and b_{1-p}) to describe fully a $TSP(a, m, b, n)$ distribution of an activity duration in order that the utilization of the TSP family by a normative expert has a lesser or preferably no substantial impact at all on the upcoming PERT analysis. In the next section we present an indirect elicitation procedure for the parameters a, b and n, utilizing directly elicited estimates (from the substantive expert) for a lower and upper quantile a_p and b_{1-p}, the mode m and *one additional quantile* (to be denoted x_s), such that

$$a_p < x_s < b_{1-p}.$$

As a side remark we note that an additional advantage of the TSP family of distributions over that of the beta family is that the TSP distributions have a greater moment ratio coverage (β_1, β_2) − at least for unimodal distributions − as compared to the beta distribution (see Figs. 3.6 and 3.7 in Chapter 3). Here, we use the notation: skewness $\beta_1 = \mu_3^2/\mu_2^3$, the kurtosis $\beta_2 = \mu_4/\mu_2^2$, $E[X] = \mu_1$ and $\mu_k = E(X - \mu_1)^k$ (for $k > 1$). Lau *et al.* (1998) note a restricted coverage of the beta family as far as β_2 is concerned. This is primarily due to the fact that the beta pdf is a smooth density function (similar to the normal or t-distributions, but with a bounded support) whereas the TSP family is "sharp" at the mode m (as is the Laplace family with unbounded support) and hence allow us to achieve sharper peakedness (i.e. higher β_2 values) than the one corresponding to the beta distribution.

4.3.3 Indirect elicitation of the parameters a, b and n

Let $X \sim TSP(a, m, b, n)$ with the pdf (4.2) and the cdf (4.3). We shall consider a more general set-up then the one suggested in Sec. 4.3.2 by allowing a separate unrestricted quantile level r in place of pre-assigned $1 - p$ for the upper quantile b_{1-p}. Thus, let a_p, b_r be the p-th and r-th percentiles of X, such that

$$a < a_p < m < b_r < b. \tag{4.27}$$

To uniquely solve for the lower bound a, upper bound b and the remaining parameter n of a TSP cdf (4.3) satisfying (4.27), we suggest elicitation of one additional quantile $x_s > m$ when $m < (a_p + b_r)/2$ or $x_s < m$ when $m \geq (a_p + b_r)/2$. Let us consider without loss of generality the case where the assessed most likely value m is less than the midpoint of a_p and b_r (obvious modifications can be made when the converse is true).

Recalling the standardized quantity $q = (m - a)/(b - a)$ (see, Eq. (4.5)) and the definition of a_p ($F(a_p|a, m, b, n) = p$), it follows from (4.3) and (4.5) that

$$a_p = a + (m - a)\sqrt[n]{\frac{p}{q}}. \tag{4.28}$$

There is no direct relation between p and q here (contrary to acceptable notation when dealing with proportions and/or the binomial distribution), except that from (4.27) and (4.5) it follows that $0 < p < q < 1$). Solving for a from (4.28), yields with (4.5)

$$a \equiv a(n, q) = \frac{a_p - m\sqrt[n]{\frac{p}{q}}}{1 - \sqrt[n]{\frac{p}{q}}} < \frac{a_p - a_p\sqrt[n]{\frac{p}{q}}}{1 - \sqrt[n]{\frac{p}{q}}} = a_p. \tag{4.29}$$

(We use the notation $a(n, q)$ instead of a to indicate that the lower bound a is a function of n and q, provided the p-th percentile a_p and the most likely value m are given.) Analogously to (4.29), we have for $m < b_r$ (using $b(n, q)$ in place of b):

$$b \equiv b(n, q) = \frac{b_r - m\sqrt[n]{\frac{1-r}{1-q}}}{1 - \sqrt[n]{\frac{1-r}{1-q}}} > \frac{b_r - b_r\sqrt[n]{\frac{1-r}{1-q}}}{1 - \sqrt[n]{\frac{1-r}{1-q}}} = b_r. \tag{4.30}$$

(Note that here we have from (4.27) and (4.5) $1 - q > 1 - r > 0$.)

Substituting $a(n, q)$ and $b(n, q)$ as given by (4.29) and (4.30) into (4.5), we arrive at the following *basic equation*

$$g(n, q) = q \tag{4.31}$$

where

$$g(n, q) = \frac{m - a(n, q)}{b(n, q) - a(n, q)} = \tag{4.32}$$

$$\frac{(m - a_p)\left(1 - \sqrt[n]{\frac{1-r}{1-q}}\right)}{(b_r - m)\left(1 - \sqrt[n]{\frac{p}{q}}\right) + (m - a_p)\left(1 - \sqrt[n]{\frac{1-r}{1-q}}\right)}$$

(Compare with Eq. (1.67) in Chapter 1.) Observe the rather "structured" relation between $g(n, q)$ and q. Indeed, from its structure it immediately follows that $0 \leq g(n, q) \leq 1$ (as it should since $g(n, q)$ also represents the probability mass to the left of the mode m). In addition, the denominator of the RHS (4.32) is "almost" a linear combination of the distances of the quantiles a_p and b_r to the mode m, where the weights are determined by the quantile probability masses p and r and the probability mass q to the left of the mode m.

In Sec. 4.3.3.3 we shall prove that — for a given value of $n > 0$ — one can numerically solve for the unique value of $q \in [p, r]$ using (4.31) and the definition of $g(n, q)$ (Eq. (4.32)) by means of for example a bisection method (see, e.g., Press *et al.* (1989)) with the starting interval $[p, r]$. Hence, relation (4.31) defines a continuous implicit function $q(n)$ with the domain $n > 0$. Next, we are able to calculate the lower bound $a\{n, q(n)\}$ (4.29) and upper bound $b\{n, q(n)\}$ (4.30) of the

$$TSP[a\{n, q(n)\}, m, b\{n, q(n)\}, n]$$

distribution given by the cdf

$$F_X(x|a\{n, q(n)\}, m, b\{n, q(n)\}, n) = \tag{4.33}$$

$$\begin{cases} g\{n, q(n)\}\left(\frac{x - a\{n, q(n)\}}{m - a\{n, q(n)\}}\right)^n, & \text{for } a\{n, q(n)\} \leq x \leq m \\ 1 - (1 - g\{n, q(n)\})\left(\frac{b\{n, q(n)\} - x}{b\{n, q(n)\} - m}\right)^n, & \text{for } m \leq x \leq b\{n, q(n)\} \end{cases}$$

where $g(n, q)$ is given by Eq. (4.32) and with pre-specified percentiles a_p and b_r satisfying

$$a\{n, q(n)\} < a_p < m < b_r < b\{n, q(n)\}.$$

Utilizing the continuity of $q(n), a\{n, q(n)\}$ and $b\{n, q(n)\}$ as a function of n one can show that the cdf (4.33) converges as $n \downarrow 0$ to a Bernoulli distribution with the probability mass

$$q(0) = \frac{m_q - a_p}{b_r - a_p} \tag{4.34}$$

at a_p and $[1 - q(0)]$ at b_r. Similarly, it follows that as $n \to \infty$, the cdf (4.33) converges to what seems to be a novel reparameterization of an asymmetric Laplace cdf with parameters a_p, m and b_{1-r}:

$$F_X(x | a_p, m, b_r) = \tag{4.35}$$

$$
\begin{cases}
q(\infty) \left\{ \frac{p}{q(\infty)} \right\}^{\frac{m-x}{m-a_p}}, & \text{for } x \leq m \\
1 - \{1 - q(\infty)\} \left\{ \frac{1-r}{1-q(\infty)} \right\}^{\frac{x-m}{b_r-m}}, & \text{for } x > m,
\end{cases}
$$

where $q(\infty)$ is the unique solution in $[p, r]$ (see Sec. 4.3.3.3 for details) that may be obtained from the equation

$$h(q) = q, \tag{4.36}$$

where

$$h(q) = \frac{(m_q - a_p) Log(\frac{1-r}{1-q})}{(b_r - m) Log(\frac{p}{q}) + (m_q - a_p) Log(\frac{1-r}{1-q})}. \tag{4.37}$$

The reader is advised to compare $g(n, q)$ (Eq. (4.32)) with (4.37). One may solve equation (4.36) utilizing for example a bisection method with starting interval $[p, r]$ for q. (An alternative standard parameterization of the asymmetric Laplace distribution is given in, e.g., Kotz *et al.* (2001).) Finally, for $n = 1$, the cdf (4.33) reduces to a uniform distribution with parameters

$$a = \frac{r a_p - p b_r}{r - p}, \quad b = \frac{(1-p) b_r - (1-r) a_p}{(r - p)}. \tag{4.38}$$

Compare (4.38) with (4.29) and (4.30) to appreciate the relation between a and a_p (or b and b_{1-r}) for various distributions.

Hence, we can render the support of the cdf (4.33) to be arbitrary large by letting $n \to \infty$, reduce it to its minimal value $[a_p, b_r]$ by letting $n \downarrow 0$ (in the case that m may also be an anti-mode), or its minimal value $[a, b]$, where a and b are given by (4.38). While the values of $n \in (0, 1)$ are not consistent with the mode m, the extent of the TSP family spanning from a Bernoulli distribution to an asymmetric Laplace one emphasizes the breath and flexibility of the family (4.2).

4.3.3.1 Description of the numerical algorithm

We are now in a position to formulate an algorithm to solve for the three remaining parameters a, b and n of a TSP distribution given a set of percentiles $\{a_p, x_s, b_{1-p}\}$ and the mode m satisfying

$$a < a_p < m < x_s < b_{1-p} < b. \tag{4.39}$$

We recommend the value 0.10 for the quantile level p and suggest the 75% or 80% quantile for s. An appeal of the 80% quantile over the 75% one is that it is reminiscent of Pareto's Law popularized in economics — also known as the 80/20 rule (see, e.g., Barabasi (2002)). The algorithm consists of the following 8 Steps :

Step 1: Set $n = 1$

Step 2: Set $r = 1 - p$, $b_r = b_{1-p}$ and solve for $q(n)$ from (4.31) and (4.32) using a bisection method with the starting interval $[p, r]$ for q. Calculate $a^* = a\{n, q(n)\}$ from (4.29) and analogously calculate $b^* = b\{n, q(n)\}$ from (4.30).

Step 3: Set $r = s$ and $b_r = x_s$. Calculate $b^\circ = b\{n, q(n)\}$ from (4.30) using $q(n)$ calculated in Step 2. If $|b^\circ - b^*| < \epsilon$ then STOP.

Step 4: If $b^\circ < b^*$ then set $n = 2n$, $n_{low} = n$, $n_{high} = 2n$ goto Step 2.

Step 5: Set $n^* = (n_{low} + n_{high})/2$.

Step 6: Set $r = 1 - p$, $b_r = b_{1-p}$ and solve for $q(n^*)$ from (4.31) and (4.32) using a bisection method with the starting interval $[p, r]$ for q. Calculate $a^* = a\{n^*, q(n^*)\}$ from (4.29) and analogously calculate $b^* = b\{n^*, q(n^*)\}$ from (4.30).

Step 7: Set $r = s$, $b_r = x_s$. Calculate $b^\circ = b\{n^*, q(n^*)\}$ from (4.30) utilizing $q(n^*)$ calculated in Step 4. If $|b^\circ - b^*| < \epsilon$ then STOP.

Step 8:   If $b° > b^*$ then $n_{high} = n^*$, else $n_{low} = n^*$. Goto Step 5.

The first steps $1 - 4$ in the algorithm above solve for a starting interval $[n_{low}, n_{high}]$ containing the solution for the parameter n of a TSP distribution with percentiles a_p, x_s and b_r and the most likely value m satisfying (4.39). The following Steps $5 - 8$ solve for n^* up to a desirable accuracy level ϵ using a bisection argument. The algorithm could be modified in an obvious manner for the case that

$$a < a_p < x_s < m < b_{1-p} < b \qquad (4.40)$$

where the mode m is to the right of the additional quantile x_s.

The proposed algorithm converges provided a set of consistent quantiles a_p, x_s, b_{1-p} satisfying (4.39) are specified together with the most likely value m. Note that it is possible to specify an inconsistent set satisfying (4.39), by setting, for example, $s = 1 - p$. On the other hand, the specification of a_p, m and b_{1-p} and the TSP family of distributions selected by a normative expert to model an activity uncertainty (or for that matter the beta family of distributions), imposes (perhaps not surprisingly) restrictions on the permissible values of the additional quantile (x_s, s). This is demonstrated in Fig. 4.8 which plots the cdf of the limiting Asymmetric Laplace distribution (4.35) with most likely value $m = 7$, $q(\infty) \approx 0.244$ and that of a uniform $[6, 11]$ distribution with support boundaries (4.38), both distributions having the common percentiles $a_p = 6.5$, $b_r = 10.5$, $p = 0.10$, $r = 1 - p = 0.90$. The boundaries for the additional quantile (x_s, s) which would be consistent with the percentiles $a_{0.1} = 6.5$, $b_{0.9} = 10.5$ and the most likely value $m = 7$, are in fact determined by the uniform and asymmetric Laplace cdf's in Fig. 4.8. Figure 4.8 also includes (for future reference) the cdf of the beta distribution depicted in Fig. 4.7 with support $[6, 15]$.

Figure 4.8 yields an alternative strategy for selecting the quantile level s of the additional quantile x_s satisfying (4.39). One could set the quantile level s, say, to a multiple of 10% that maximizes the allowable quantile range of x_s as specified by the inverse of the uniform distribution with support $[a, b]$ where a and b are given by (4.38) and the inverse of the limiting AL distribution (4.35) given by

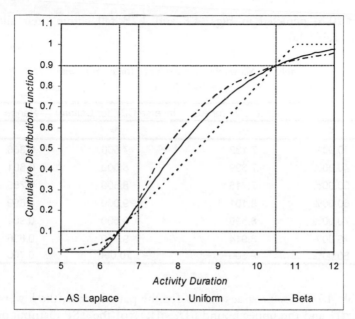

Fig. 4.8 CDF's of asymmetric Laplace distribution (4.35) satisfying $m = 7$, a uniform distribution with the bounds $a = 6$, $b = 11$ given by (4.38) and that of a beta distribution with parameters $a = 6$, $b = 15$, $\alpha = 1.384, \beta = 4.071$ (4.1). All three distributions have the common percentiles $a_p = 6.5$, $b_r = 10.5$, $p = 0.10$ and $r = 1 - p = 0.90$.

$$F_X^{-1}(y|a_p, m, b_r) =
\begin{cases}
m - (m - a_p)\dfrac{Log(\frac{y}{q(\infty)})}{Log(\frac{p}{q(\infty)})}, & \text{for } y \le q(\infty) \\[3mm]
x = m + (b_r - m)\dfrac{Log(\frac{1-y}{1-q(\infty)})}{Log(\frac{1-r}{1-q(\infty)})}, & \text{for } y > q(\infty).
\end{cases}$$

Table 4.6 provides some allowable quantile ranges for x_s for different values of the quantile level s derived from the distributions presented in Fig. 4.8. From Table 4.6 it follows that in this particular example, the 70% quantile yields the largest allowable quantile range 0.901 for the additional quantile x_s and almost the same range is attained for the 60% quantile. Note that, the percentiles $x_{0.70} = 8.952$, $x_{0.75} = 9.250$ and $x_{0.80} = 9.588$ of the beta distribution (with the mode $m = 7$, $a_{0.10} = 6.5$ and

$m_{0.90} = 10.5$) depicted in Figs. 4.8 and 4.7B fall within their allowable ranges in Table 4.6.

Table 4.6 Allowable quantiles range for $x_s > m$ with
$a_p = 6.5, m = 7\ b_r = 10.5, p = 0.10, r = 1 - p = 0.90.$

	Inverse CDF for AL	Inverse CDF for Uniform	Range
s	x_s	x_s	
30.00%	7.132	7.500	0.368
40.00%	7.399	8.000	0.601
50.00%	7.715	8.500	0.785
60.00%	8.101	9.000	0.899
70.00%	8.599	9.500	0.901
75.00%	8.914	9.750	0.836
80.00%	9.300	10.000	0.700

Table 4.7 below provides values for the parameter n, the lower bound $a\{n, q(n)\}$ and the upper bound $b\{n, q(n)\}$ of the TSP distributions that follow from the algorithm described above, by using the percentiles $a_{0.1} = 6.5$, $b_{0.9} = 10.5$, the mode $m = 7$ and the values x_s for an additional quantile of the beta distribution in Figs. 4.8 and 4.7, for $s = 0.70, 0.75$ and 0.80. Moreover, Table 4.7, compares this beta distribution and the TSP distributions in terms of their mean values, variances, skewnesses and kurtoses. For the case presented in Table 4.7, a slight advantage can perhaps be assigned to the TSP distribution in fourth column (TSP C), since its values align closer to those of its beta counterpart (presented in the first column).

Figure 4.9 plots the cdf and the pdf of the beta distribution in Table 6 (and Figs. 4.8 and 4.7) and that of the TSP C distribution mentioned above. In Fig. 4.9B, the common percentiles of the beta and TSP cdf's are indicated by dotted lines. The pdf's and cdf's associated with the second (TSP A) and the third (TSP B) columns in Table 4.7 are not presented in Fig. 4.9, but visually are almost indistinguishable from those for the TSP C case represented in Fig. 4.9. (Please keep in mind the values of skewness and kurtosis for the standard normal distribution are 0 and 3, respectively.)

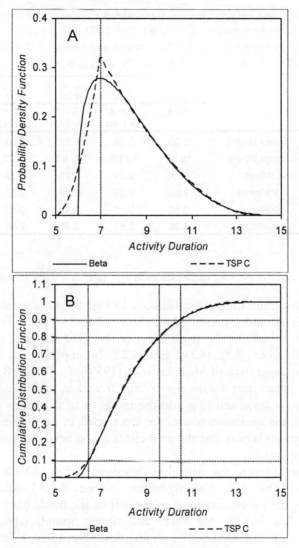

Fig. 4.9 PDF and CDF of a beta [6, 15] distribution with parameters $\alpha = 1.384, \beta = 4.071$ and TSP C distribution (presented in the fourth column of Table 4.7) both with the same percentiles $a_{0.10} = 6.5, x_{0.80} = 9.588, b_{0.90} = 10.5$ and the most likely value $m = 7$.

Table 4.7 Lower and upper bounds, mean, variance, skewness and kurtosis of a beta $[6, 15]$ distribution with parameters $\alpha = 1.384, \beta = 4.071$ and TSP distributions (A,B,C), all with common percentiles $a_{0.10} = 6.5, b_{0.90} = 10.5$ and mode $m = 7$. A: $x_{0.70} = 8.952$, $n = 2.682$; B : $x_{0.75} = 9.250$, $n = 2.729$; C: $x_{0.80} = 9.588$, $n = 2.782$.

	Beta	$x_{0.70} = 8.952$ TSP A $n = 2.682$	$x_{0.75} = 9.250$ TSP B $n = 2.729$	$x_{0.80} = 9.588$ TSP C $n = 2.782$
Lower Bound	6.00	5.09	5.06	5.03
Upper Bound	15.00	13.56	13.63	13.72
Mean	8.28	8.26	8.26	8.26
Variance	2.38	2.37	2.37	2.38
Skewness	0.60	0.40	0.42	0.43
Kurtosis	3.08	2.82	2.85	2.88

4.3.3.2 Assessment of the effect of the elicitation procedure in a PERT example

Consider once more the project network in Fig. 4.4, its activities and their values for a, m, and b presented in Table 4.2 in Sec. 4.3.2. Table 4.2 also provides the values for the parameters α and β of the beta distributions which follow from (4.1), (4.21) and (4.22) by applying the methods of moments (by suggestion of Malcolm *et al.* (1959)). In Table 4.8, we present again these parameters values for α and β and in addition provide the percentiles $a_{0.10}, x_{0.80}$ and $b_{0.90}$ and the mode m of these beta pdf's to test the effect of the elicitation procedure developed in the beginning of Sec. 4.3.3. Our main claim is that the new elicitation procedure is advantageous in several respects.

Next, we execute the algorithm described in Sec. 4.3.3.1 for the 18 activities in Table 4.2. The algorithm in Sec. 4.3.3.1 solves for the parameter n, the probability mass to the left of the mode $q(n)$ (the unique solution of Eq. (4.31)), the lower and upper bounds $a\{n, q(n)\}$ and $b\{n, q(n)\}$ of TSP distributions with identical percentiles and modal values as those presented in Table 4.8. The resulting values for $n, q(n)$, $a\{n, q(n)\}$ and $b\{n, q\{n\}\}$ are displayed in Table 4.9.

We now generate the cdf of the completion time distribution of the project presented in Fig. 4.4 (using the Monte Carlo technique involving 25000 samples) via the beta distributions in Table 4.8, the TSP distributions in Table 4.9, triangular distributions with parameters a, m and b in Table

Table 4.8 Parameters α and β of beta distribution for activity durations of the project network in Fig. 4.4 together with their percentiles $a_{0.10}$, $x_{0.80}$ and $b_{0.90}$ and modes m.

ID	α	β	$a_{0.1}$	m	$x_{0.8}$	$b_{0.9}$
1	2.94	4.62	23.58	25.14	26.83	27.56
2	3.23	4.52	36.61	38.10	39.52	40.15
3	1.32	4.21	20.35	21.51	29.57	32.39
4	1.34	4.24	6.45	6.86	9.42	10.32
5	1.56	4.40	23.46	23.99	25.82	26.50
6	3.40	4.44	16.16	18.11	19.82	20.59
7	3.74	4.22	11.75	14.05	15.81	16.62
8	2.33	4.67	6.01	7.13	8.83	9.52
9	2.59	4.67	27.03	28.12	29.52	30.11
10	3.91	4.08	28.32	35.06	39.81	42.05
11	2.70	4.66	28.55	30.17	32.13	32.96
12	2.13	4.64	5.99	7.13	9.14	9.94
13	1.97	4.59	4.49	5.06	6.22	6.67
14	1.73	4.49	6.47	7.04	8.52	9.09
15	1.56	4.40	9.46	9.99	11.82	12.50
16	1.34	4.24	6.45	6.86	9.42	10.32
17	1.56	4.40	19.46	19.99	21.82	22.50
18	1.84	4.54	13.96	15.10	17.74	18.76

Table 4.9 The value of the parameter n, probability mass to the left of the mode $q(n)$ and lower and upper bounds $a\{n, q(n)\}$ and $b\{n, q(n)\}$ of TSP distributions that follow from the algorithm in Sec 4.3.3.1 utilizing the percentiles of the beta distributions presented in Table 4.8 and their modes.

ID	n	$q(n)$	$a\{n,q(n)\}$	$b\{n,q(n)\}$
1	2.14	42.4%	21.97	29.46
2	2.04	44.5%	35.22	41.70
3	2.95	20.2%	16.06	43.01
4	2.92	21.1%	5.04	13.68
5	2.73	26.6%	22.23	28.84
6	1.98	45.6%	14.48	22.43
7	1.86	48.0%	10.01	18.43
8	2.35	37.3%	4.51	11.55
9	2.25	39.7%	25.73	31.75
10	1.80	49.3%	23.61	46.82
11	2.22	40.5%	26.72	35.23
12	2.43	35.2%	4.29	12.36
13	2.51	33.3%	3.55	8.10
14	2.63	29.7%	5.38	10.95
15	2.73	26.6%	8.23	14.84
16	2.92	21.1%	5.04	13.68
17	2.73	26.6%	18.23	24.84
18	2.57	31.4%	11.92	22.04

4.2 and $TSP(a, m, b, 5)$ distribution with parameters values $n = 5$ which ensures the equality of $E[T]$ in (4.21) and (4.6) (here $n - 1 = 4$ and $n + 1 = 6$). The resulting cdf's are depicted in Fig. 4.10, where "TSP(varying n)" indicates the case of the completion time distribution of the Project in Fig. 4.4, associated with Table 4.9.

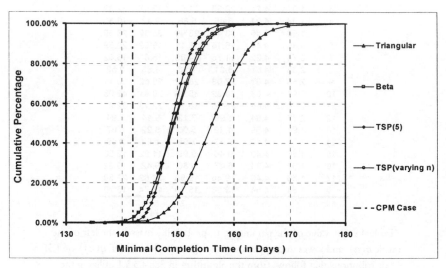

Fig. 4.10 Comparison of cumulative distribution functions of
the completion time for the project in Fig. 4.4 in Sec. 4.3.2..

Figure 4.10 seems to be worthy of a scrupulous examination being a culmination of various theoretical and numerical derivations in this chapter. We shall present a number of indicative features. First note that Fig. 4.10 reconfirms the analysis in Table 4.3. Namely, the triangular case (suggested by D. Johnson (1997)) has quite a substantial mean shift as compared to the beta case (suggested by Malcolm *et al.* (1959)) as well as a shallow slope indicating a larger variance. The TSP(5) case suggested in Van Dorp and Kotz (2002b) when compared with this beta case is similar in their mean values (see also, Table 4.3), but has a steeper slope and hence a smaller variance. The project completion time 142 days that follow from the CPM case (using only the most likely values m) is represented by a semi-dotted vertical line in Fig. 4.10. Considering that the values of the modes m being less than the midpoint $(a + b)/2$ for all 18 activities in Table 4.2, we observe from Fig. 4.10 that the probability of meeting the completion time of 142 days is less than 1%. Although the skewness of the activity distributions in Table 4.2 may be somewhat overstated (for illustrative purposes), a case could be made that skewness towards the lower bound does tends to appear in assessed activity time distributions due to a motivational bias of the substantive expert. The latter points to the necessity of conducting a project risk analysis (such as PERT) instead of the basic CPM analysis in the first place.

Finally, and perhaps most importantly, the cdf involving the "TSP(varying n)" case associated with Table 4.9 differs only slightly from the completion time cdf associated with beta distributions. The values for the mean, standard deviation, minimum and maximum observed values of the minimal completion time of the Project Network are compared in Table 4.10 for these two cases. Note that these values are in agreement with the earlier observations concerning their cdf's in Fig. 4.10. Hence, − and here comes our main conclusion − *by eliciting an additional quantile x_s satisfying (4.39) or (4.40) from a substantive expert, a much lesser effect (compared to the ones resulting in the earlier mentioned PERT "controversy") is observed in the completion time of a project for the choice of a normative expert to model the uncertainty in an activity duration by a TSP or beta distribution.*

In the next section we shall elaborate on some mathematical details regarding the development of the numerical routine in Sec. 4.3.3.1 which solves for the lower and upper bounds a and b, and the shape parameter n of a $TSP(a, m, b, n)$ pdf (4.2) given a lower quantile estimate $\widehat{a}_p$, and upper quantile estimate $\widehat{b}_{1-p}$, a most likely estimate $\widehat{m}$ and an estimate of the additional quantile (to be denoted $\widehat{x}_s$), such that $\widehat{a}_p < \widehat{x}_s < \widehat{b}_{1-p}$, when these estimates are elicited from a substantive expert.

Table 4.10 Mean and Standard Deviation of the Project Completion Time
Distribution using Beta distribution (suggested by Malcolm *et al.* (1959))
and the TSP (varying n) distribution associated with Table 4.9.

	Mean	Standard Deviation	Min	Max
Beta	150.5	4.7	134.4	169.4
TSP(Varying n)	150.3	4.4	134.9	167.8

4.3.3.3 Some mathematical details regarding the algorithm in Sec. 4.3.3.1

In this section we shall dwell on the implicit function $q(n)$, which is an important new concept introduced in this chapter. It plays a pivotal role in the algorithm described in Sec. 4.3.3.1. Let $X \sim TSP(a, m, b, n)$ with pdf (4.2) and cdf (4.3). From inequality (4.27) and an expression for the probability mass q (4.5) to the left of the mode m, we have that the relations

$$0 < p < q < r < 1 \Rightarrow 0 < \frac{p}{q} < 1 \quad \text{and} \quad 0 < \frac{1-r}{1-q} < 1 \quad (4.41)$$

are valid.

Taking partial derivatives with respect to n and q of the function $a(n, q)$ given by (4.29) (corresponding to the lower boundary) we obtain

$$\frac{\partial}{\partial n} a(n, q) = \frac{1}{n^2} Log\left(\frac{p}{q}\right) \sqrt[n]{\frac{p}{q}} \, \frac{m - a_p}{\{1 - \sqrt[n]{\frac{p}{q}}\}^2} < 0,$$

and

$$\frac{\partial a(n, q)}{\partial q} = \frac{1}{nq} \sqrt[n]{\frac{p}{q}} \, \frac{m - a_p}{(1 - \sqrt[n]{\frac{p}{q}})^2} > 0. \quad (4.42)$$

Hence, $a(n, q)$ is a strictly decreasing (increasing) function of n (of q). In addition, from (4.41), (4.27) and (4.29) it follows that

$$a(n, q) \to -\infty \text{ as } n \to \infty \quad (q \downarrow p) \quad (4.43)$$

for the values of $q \in (p, r)$ (of $n > 0$). Analogously, taking partial derivatives with respect to n and q of the upper bound function $b(n, q)$ (4.30) results in

$$\frac{\partial b(n, q)}{\partial n} = -\frac{1}{n^2} Log\left(\frac{1-r}{1-q}\right) \sqrt[n]{\frac{1-r}{1-q}} \, \frac{b_r - m}{\{1 - \sqrt[n]{\frac{1-r}{1-q}}\}^2} > 0,$$

and

$$\frac{\partial b(n, q)}{\partial q} = \frac{1}{n(1-q)} \sqrt[n]{\frac{1-r}{1-q}} \; \frac{b_r - m}{\{1 - \sqrt[n]{\frac{1-r}{1-q}}\}^2} > 0. \qquad (4.44)$$

Hence, the upper bound $b(n, q)$ is a strictly increasing function of both n and q. In addition, from (4.41), (4.27) and (4.30) it follows that

$$b(n, q) \to \infty \text{ as } n \to \infty \quad (q \uparrow r) \qquad (4.45)$$

for values of $q \in (p, r)$ (of $n > 0$).

The basic equation $g(n, q) = q$ (4.31) where $g(n, q)$ is defined in (4.32) plays a pivotal role in the arguments leading to determination of the parameters of a $TSP(a, m, b, n)$ pdf via the algorithm developed in Sec. 4.3.3.1. Hence, this equation is the key for the newly proposed indirect elicitation procedure for the parameters a, b and n given elicited values for a lower and upper quantile a_p and b_{1-p}, the mode m and additional quantile x_s all satisfying (4.39) or (4.40). The theorem below provides details concerning the existence of a unique solution to this equation.

Theorem 4.1 : *The equation* $g(n, q) = q$ *(4.31) has a unique solution* $q(n) \in (p, r)$ *for every fixed value* $n > 0$, *where* $g(n, q)$ *is defined in (4.32), and moreover the resulting implicit function* $q(n)$ *is continuous.*

Proof : Let $n > 0$ be a fixed value. Substituting $q = p$ into (4.32) we have

$$g(n, p) = 1. \qquad (4.46)$$

Substituting $q = r$ into (4.32) we have

$$g(n, r) = 0. \qquad (4.47)$$

From the continuity of (4.32) (as a function of q for fixed $n > 0$) on $q \in (p, r) \subset [0, 1]$ we conclude that a solution $q^* \in (p, r)$ of (4.31) exists for any value of $n > 0$. Uniqueness of q^* would follow if the function $g(n, q)$ is a non-increasing as a function of $q \in (p, r)$ (as will be shown below). Note that the condition of $g(n, q)$ being non-increasing — as opposed to strictly decreasing — in q is sufficient since we are solving here for a *root* of the equation $g(n, q) = q$. Moreover, this is equivalent to showing that the reciprocal of $g(n, q)$:

$$\{g(n,q)\}^{-1} = \frac{b(q,n) - a(q,n)}{m - a(q,n)} = 1 + \frac{b(q,n) - m}{m_q - a(q,n)}$$

is a non-decreasing function of $q \in (p,r)$. The latter follows immediately from the fact that $a(q,n)$ and $b(q,n)$ are strictly increasing functions of $q \in (p,r)$ (see, (4.42) and (4.44), respectively). Hence, one can write $q^{\bullet} = q(n)$, where $q(n)$ is the unique solution to (4.31) for a fixed value of $n > 0$. Continuity of the implicit function $q(n)$ then follows from the classical implicit function theorem (see, e.g., Krantz 2002) and the continuity of the $g(n,q)$ in both parameters n and q for $n > 0$ and $q \in (p,r)$. $\square$

Note that from (4.46) (from (4.47)) it follows that the function $g(n,q)$ is right-continuous (left-continuous) as a function of q for a fixed value $n > 0$ at $q = p$ (at $q = r$) since $g(n,q) \to 1$ $(g(n,q) \to 0)$ as $q \downarrow p$ (as $q \uparrow r$) for any value $n > 0$. Hence, the unique solution $q(n)$ of the Eq. (4.31) may be obtained either by using standard bisection methods with a (closed) starting interval $[p,r]$ for $q(n)$ or by employing a root finding algorithm such as GOALSEEK available in Microsoft EXCEL. Observe, however that the right-continuity at $q = p$ (at $q = r$) does not hold for the function $a(n,q)$ (the function $b(n,q)$) defined in (4.29) (in (4.30)).

Returning to the implicit function $q(n)$ we shall now investigate its behavior and consider the three cases: $n \downarrow 0$, $n = 1$ (the uniform case) and $n \to \infty$.

A) Case $n \downarrow 0$: From (4.41), (4.29) and (4.30) it follows that

$$\lim_{n \downarrow 0} \sqrt[n]{\frac{p}{q}} = 0 \ \Rightarrow \ \lim_{n \downarrow 0} a(n,q) = a_p, \tag{4.48}$$

$$\lim_{n \downarrow 0} \sqrt[n]{\frac{1-r}{1-q}} = 0 \ \Rightarrow \ \lim_{n \downarrow 0} b(n,q) = b_r. \tag{4.49}$$

Hence, from the basic equation (4.31) one obtains

$$q(0) = \lim_{n \downarrow 0} q(n) = \frac{m_q - a_p}{b_r - a_p}. \tag{4.50}$$

B) Case $n = 1$: In this case the TSP cdf (4.3) reduces to a uniform distribution and we may solve directly for its lower bound a and the upper bound b from the specified percentiles a_p and b_r utilizing properties of similar triangles and the fact the cdf $F_X(x|a, m_q, b, 1)$ of a uniform variable $X \sim TSP(a, m_q, b, 1)$ is a linear function between the extreme coordinates $(a, 0)$ and $(b, 1)$ crossing through (a_p, p) and (b_r, r) (see Fig. 4.11).

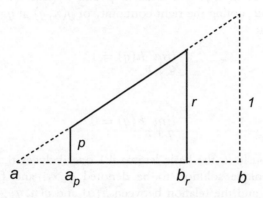

Fig. 4.11 Similar triangles used to solve for the lower bound a and upper bound b of a uniform distribution on $[a, b]$ with specified percentiles a_p and b_r.

From Fig. 4.11 and similarity of the triangles we have

$$\frac{a_p - a}{p} = \frac{b_r - a}{r} \Rightarrow a = \frac{r a_p - p b_r}{r - p} \qquad (4.51)$$

and using (4.51)

$$\frac{b_r - a}{r} = b - a \Rightarrow b = \frac{(1 - p)b_r - (1 - r)a_p}{(r - p)}. \qquad (4.52)$$

Substituting (4.51) and (4.52) into (4.31) one arrives at

$$q(1) = \frac{p b_r + (r - p)m_q - r a_p}{b_r - a_p} \qquad (4.53)$$

C) Case $n \to \infty$: From the basic interrelation between $g(n, q)$ and q (4.32), the ordering of a, a_p, m, b_r and b as given by (4.27), and the relation between p, q and r (4.41), recalling the definition of $h(q)$ (4.37) (a logarithmic counter part of $g(n, q)$), it is easy to verify that

$$\lim_{n \to \infty} g(n, q) = h(q). \tag{4.54}$$

By the continuity of $g(n, q)$ and the fact that $g(n, q)$ is non-increasing as a function of q it follows that the function $h(q)$ is also non-increasing. From the properties of $g(n, q)$ evaluated at p and r (Eqs. (4.46) and (4.47), respectively) and noting the right continuity of $g(n, q)$ at $q = p$, it follows immediately that

$$\lim_{q \downarrow p} h(q) = 1$$

and

$$\lim_{q \uparrow r} h(q) = 0.$$

Hence, analogously to the basic lemma 4.1 one deduces that the equation (4.36) has a unique solution to be denoted $q(\infty)$ and from the basic equation (4.31) and the relation between $h(q)$ and $g(n, q)$ (4.54) it follows that

$$\lim_{n \to \infty} q(n) = q(\infty).$$

The value of $q(\infty)$ may be determined from the equation $h(q) = q$ (4.36) and the definition of $h(q)$ (4.37) by using a standard bisection method with the starting interval $[p, r]$ for $q(\infty)$.

Figure 4.12 provides an example of the function $q(n)$ for the case $p = 0.10$, $r = 1 - p = 0.90$, $a_p = 6.5$, $m = 7$ and $b_r = 10.5$. Observe that these values coincide with the values used in Fig. 4.7 in Sec. 4.3.2.1. In Fig. 4.12 the values of $q(0)$ (see, Eq. (4.50)), $q(1)$ (see, Eq. (4.53)) and $q(\infty) \approx 0.244$ (calculated from (4.36) and (4.37) using a bisection method) are presented. In addition, Fig. 4.12 depicts the value of $q(2)$ (of $q(7)$) associated with the Triangular (the TSP(7)) distribution in Fig. 4.7 calculated from (4.31) and (4.32) also employing a bisection method. Note a painfully slow increase in $q(n)$ in Fig. 4.12 as a function of n ($n \geq 3$). Figure 4.13 plots the lower bound function $a\{n, q(n)\}$ (Eq. (4.29)) and upper bound function $b\{n, q(n)\}$ (Eq. (4.30)) associated with $q(n)$ in Fig. 4.12. Observe an almost linear behavior of both the functions $a\{n, q(n)\}$ and $b\{n, q(n)\}$ as a function of n.

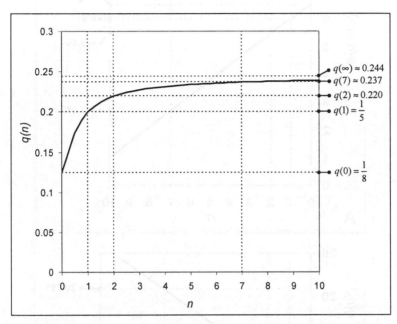

Fig. 4.12 Graph of the implicit function $q(n)$ satisfying (4.31) for the case $p = 0.10$, $r = 1 - p = 0.90$, $a_p = 6.5$, $m_{q(n)} = 7$ and $b_r = 10.5$.

Utilizing continuity of $q(n), a\{n, q(n)\}$ and $b\{n, q(n)\}$ as a function of n, limiting relations (4.48) and (4.49)) one can show that the TSP cdf (4.33) converges as $n \downarrow 0$ to a Bernoulli distribution with a probability mass $q(0)$ (4.50) at a_p and $\{1 - q(0)\}$ at b_r. Similarly, utilizing (4.43) and (4.45) it could be verified that the TSP cdf (4.33) converges to the Laplace cdf (4.35) with the pdf

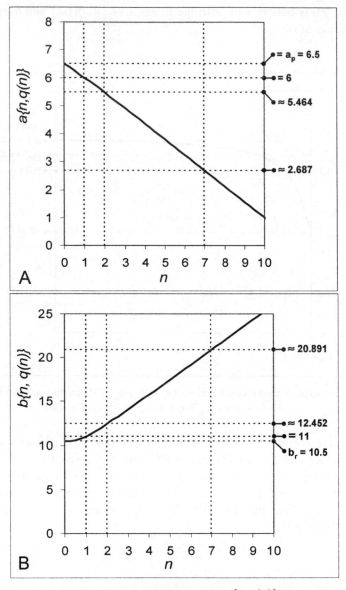

Fig. 4.13. Graphs of A: the lower bound function $a\{n, q(n)\}$ (Eq. (4.29)) and B: the upper bound function $b\{n, q(n)\}$ (Eq. (4.30)), where $q(n)$ is the implicit function depicted in Fig. 4.12.

$$f_X(x|a_p, m, b_r) = \qquad\qquad\qquad\qquad\qquad (4.55)$$

$$\begin{cases} q(\infty)\mathcal{A}Exp\left\{-\mathcal{A}(m-x)\right\}, & \text{for } x \le m \\[2mm] \{1 - q(\infty)\}\mathcal{B}Exp\left\{-\mathcal{B}(x-m)\right\}, & \text{for } x > m \end{cases}$$

where the coefficients are

$$\mathcal{A} = \frac{Ln\left\{\frac{q(\infty)}{p}\right\}}{m - a_p} \text{ and } \mathcal{B} = \frac{Ln\left\{\frac{1-q(\infty)}{1-r}\right\}}{b_r - m}.$$

The structural form of the cdf (4.33) and the pdf (4.55) is recognized to be a somewhat unexpected new "unorthodox" reparameterization of the asymmetric Laplace (AL) distribution (see, e.g., Kotz *et al.* (2001)) with a pre-specified mode m, lower quantile a_p and upper quantile b_r.

4.4 Concluding Remarks

A new four-parameter family of $TSP(a, m, b, n)$ distributions has been proposed which possesses some attractive properties, especially those related to the meaning of its parameters and the structure of its expected value as a function of parameters, as well as an instructive and algorithmically quite straightforward new maximum likelihood estimation procedure. The family of TSP distributions naturally extends the three-parameter triangular distributions. The new four-parameter $TSP(a, m, b, n)$ distribution seems to be a useful and a more flexible alternative to the four-parameter beta distribution than the triangular distribution, specifically in, but not limited to, PERT applications described in this chapter. It is our hope that the introduction of the proposed distribution into statistical practice will assist in the basic goals of applied statistical work.

In addition, we propose a novel method to solve for the parameters a, b and n of a TSP distribution from a lower quantile estimate $\widehat{a}_p$, most likely estimate $\widehat{m}$, an upper-quantile estimate $\widehat{b}_{1-p}$ and one additional quantile $\widehat{x}_s$ (to be determined by a substantive expert) such that $\widehat{a}_p < \widehat{x}_s < \widehat{b}_{1-p}$. The method utilizes an algorithm developed in Sec. 4.3.3.1, by first setting $m = \widehat{m}$ and next successively solving a single non-linear equation in the unknown probability mass $q = (m-a)/(b-a)$ (Eq. (4.5)) to the left of the mode m. Section 4.3.3.3 discusses some mathematical features exhibited in the development of this algorithm. In addition, we have shown a much lesser effect on the completion time distribution of the PERT example (in Fig. 4.4) when a normative expert models activity duration via a beta or TSP distributions given the estimates $\widehat{a}_p$, $\widehat{x}_s$, $\widehat{b}_{1-p}$, $\widehat{m}$ than it was previously encountered in applications (which resulted in a PERT "controversy" for several decades.)

While the decision to model random variables via beta, triangular or TSP distributions based on a lower bound estimate $\widehat{a}$, most likely estimate $\widehat{m}$, and upper bound estimate $\widehat{b}$ has received quite some attention in the academic oriented literature in the context of PERT, it does not seem to attract similar attention in the context of Monte Carlo methods in general or discrete event simulation in particular. In fact, a recent college text book on discrete event simulation (Altiok and Melamed (2001)) containing the popular simulation package ARENA (by Rockwell Software Inc.) explicitly recommends the triangular distribution when the underlying distribution is unknown, but a minimal value $\widehat{a}$, some maximal value $\widehat{b}$ and a most likely value $\widehat{m}$ are available. Some discrete event simulation text books (Kelton *et al.* (2002), Banks *et al.* (2000)) define a triangular uncertainty model when only the minimum, most likely and maximum values of the distribution are known, but do not mention that other families of bounded distributions (such as the beta or TSP family) may reflect the same information albeit in a different manner (that may substantially affect the results of the simulation analysis). (In fairness it should be mentioned that the TSP distribution was introduced only in 2001, however, the beta distribution has been popular and widespread for some 100 years.)

From the analyses in this chapter it follows that at least the beta and TSP models should also be considered in such instances. Furthermore, it is our opinion that the choice of a triangular distribution by a normative expert to model the random input variable in a simulation based solely on the estimates $\widehat{a}$, $\widehat{b}$ and $\widehat{m}$ may affect the simulation analysis to the extent comparable to that in the PERT example discussed above. We also believe that the material in this chapter indicates that the latter (simplified) decision by a normative expert could perhaps be somewhat rushed and that an expert should definitely consider the elicitation of specific additional information prior to implementing such a modeling choice.

Finally, we would like to comment on a phenomenon that has shown to have a similar effect on the project completion time distribution in the PERT context as the uncertainty model for activity durations. We have in mind the existence of almost inevitable statistical dependence amongst the input distributions — even in the case when this dependence is mild (see, e.g., van Dorp (2004)). Hence, it would seem that the topic of modeling statistical dependence in simulation analysis (in PERT and in more general simulation contexts) deserves comparable attention in simulation text books as the topic of selecting an uncertainty model. To the best of our knowledge this is not yet the case. While the topic is touched on in Law and Kelton (1991) and Altiok and Melamed (2001), it does not appear to be mentioned in the simulation text books Kelton *et al.* (2002) and Banks *et al.* (2000), which may give the misleading impression that uncertainty modeling requires solely specification of marginal distributions. While it is perhaps possible to specify dependence in standard uncertainty analysis software — such as @Risk (developed by the Palisade Corporation) and Crystal Ball (developed by Decision Engineering) and even simulation packages such as ARENA mentioned above — one should not rely on packages of this kind to educate students about statistical dependence. Modeling of statistical dependence should in fact be an integral part of the simulation and uncertainty analysis curriculum.

Chapter 5

The Generalized Trapezoidal Distribution

Geometric concepts and arguments as a rule do not provide important tools in classical probabilistic models and distributions. The father of the modern systematic classification of distributions K. Pearson (1857-1936) did not utilize them when developing his famous probability curves. His brilliant successor R.A. Fisher (1890-1962) was more inclined to use geometry in particular when developing properties of normal distributions. Only quite recently, the popularity of fractals have increased interest and application of geometry in the area of univariate statistical models. Our investigations of triangular distributions in Chapter 1 seems to naturally lead us to the trapezoidal ones (discussed briefly in Chapter 2) — which can be viewed as a direct geometric extension, albeit (seemingly) devoid of a probabilistic motivation. Moreover, it is with some hesitation that we approach a generalization of the four-parameter trapezoidal distribution (which will contain seven parameters), bearing in mind that any generalization of a distribution *inevitably* involves introduction of additional parameters. In such a case, the inconsistent and wasteful situation of over-parameterization may arise. Fortunately, the parameters of our generalized trapezoidal distribution (GTD) are sharply and distinctly defined without an overlap or interdependence. We present the construction and basic properties of GTD distributions of an arbitrary form defined on a compact (bounded) set by concatenating in a continuous manner three pdf's with bounded support using a modified mixture technique. These three distributional components could represent the growth, stability and decline stages of the likelihood of a certain physical or mental phenomenon.

5.1 Illustrative Example

As it was mentioned earlier in Chapter 2, Sec. 2.2 trapezoidal distributions have been advocated in risk analysis problems by Pouliquen already in 1970 and more recently by Powell and Wilson (1997) and Garvey (2000). They have also found application as membership functions in the fuzzy set theory (see, e.g., Chen and Hwang (1992)). Another domain for applications of trapezoidal distributions is the applied physics arena (see, e.g. Davis and Sorenson (1969), Nakao and Iwaki (2000), Sentenac *et al.* (2000), Straaijer and De Jager (2000). Specifically, in the context of nuclear engineering, uniform and trapezoidal distribution have been assumed as models for the observed axial distributions for burnup credit calculations (see Wagner and DeHart (2000) and Neuber (2000) for a comprehensive description). These distributions are of relevance to burnup credit criticality safety analyses for pressurized-water-reactor (PWR) fuels. Table 5.1 and Fig. 5.1 (adapted from Wagner and DeHart (2000)) depict the actual data and two profiles of axial normalized burnup likelihood versus percent axial height (using an interpolation between the observed data points). Apparently, the uniform distribution has been shown to be suitably conservative only for low burnups, but not when burnup increases (see, Wagner and DeHart (2000)) and the use of trapezoidal distributions tend to result in more conservative criticality safety analyses (see, Neuber (2000)). The explicit modeling of axial burnup distributions is becoming an important and timely research topic in nuclear engineering (see Parks *et al.* (2000)).

Motivated by the structural form of the profiles in Fig. 5.1, we shall strive for a continuous generalization of the trapezoidal distribution where the growth and decay stages may exhibit a nonlinear convex or concave behavior and the density values $f_X(\cdot)$ between b and c are not necessarily the same, but follow a linear form. In the proposed generalization, a *boundary ratio parameter* $\alpha > 0$ is introduced such that $f_X(b) = \alpha f_X(c)$. These generalized trapezoidal distributions inherit the four basic trapezoidal parameters a, b, c and d (see Eq. (2.22) in Chapter 2 and Fig. 2.6) and require, for complete specification, two additional parameters n_1 and n_3 specifying the growth rate and decay rates in the first and third stages of the distribution and also the above mentioned boundary ratio parameter α. An attractive feature of this generalized trapezoidal distribution is its flexibility which allows us, *inter alia*, to appropriately mimic a great variety of growth and decay behaviors.

Figure 5.2 depicts two members in the generalized trapezoidal family that closely follow the axial distribution profiles presented in Fig 5.1. From Fig. 5.2 it follows that the density function of a generalized trapezoidal distribution (to be discussed below) may well be applicable for modeling

Table 5.1 Typical profiles of observed normalized burnup likelihoods
as a function of % axial height in PWR. (Source: Wagner and DeHart (2000).)

	Normalized Burnup	
Axial Height %	Profile 1	Profile 2
2.78%	0.652	0.649
8.33%	0.967	1.044
13.89%	1.074	1.208
19.44%	1.103	1.215
25.00%	1.108	1.214
30.56%	1.106	1.208
36.11%	1.102	1.197
41.69%	1.097	1.189
47.22%	1.094	1.188
57.80%	1.094	1.192
58.33%	1.095	1.195
63.89%	1.096	1.190
69.44%	1.095	1.156
75.00%	1.086	1.022
80.56%	1.059	0.756
86.11%	0.971	0.614
91.67%	0.738	0.481
97.22%	0.462	0.284

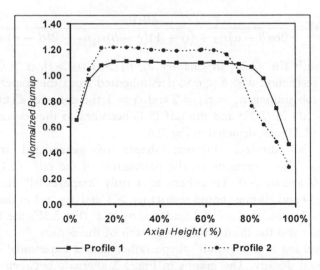

Fig. 5.1 Axial distributions in PWR for the data in Table 5.1.
(Source: Wagner and DeHart (2000).)

149

axial burnup distribution profiles. Note especially the graph B, where the decline in the central part is closely tracked. Applications to reliability and risk analysis also become more realistic by replacing the linear parts of the trapezoidal pdf defined by Eq. (2.22) in Chapter 2 with a more general power function.

5.2 The Functional Form of the Generalized Trapezoidal Density

We begin by providing the functional form of the generalized trapezoidal distribution followed by a discussion detailing its construction. The pdf of a generalized trapezoidal distribution is given by

$$
f_X(x|\Theta) = C(\Theta) \times
\begin{cases}
\alpha \left(\frac{x-a}{b-a} \right)^{n_1-1}, & \text{for } a \leq x < b \\
\left\{ (\alpha - 1)\frac{c-x}{c-b} + 1 \right\}, & \text{for } b \leq x < c \quad (5.1) \\
\left(\frac{d-x}{d-c} \right)^{n_3-1}, & \text{for } c \leq x < d
\end{cases}
$$

where $\Theta = \{a, b, c, d, n_1, n_3, \alpha\}$, the multiplier

$$
C(\Theta) = \frac{2n_1 n_3}{2\alpha(b-a)n_3 + (\alpha+1)(c-b)n_1 n_3 + 2(d-c)n_1}, \quad (5.2)
$$

(compare with Eq. (2.23) in Chapter 2), $n_1 > 0, n_3 > 0, \alpha > 0$ and the parameter restriction $a < b < c < d$ is inherited from the trapezoidal pdf (2.22). By substituting $n_1 = n_3 = 2$ and $\alpha = 1$ the constant $C(\Theta)$ reduces to $C(a, b, c, d)$ in (2.23) and the pdf (5.1) becomes to the trapezoidal pdf defined by (2.22) and depicted in Fig. 2.6.

   Figure 5.3 displays different shapes of generalized trapezoidal distributions. The conditions on the parameters of the pdf (5.1) stipulate that $n_1 > 0$ and $n_3 > 0$. To adhere to a truly "trapezoidal" shape (Figs. 5.3A, B, C, D and E) one should restrict $n_1 > 1$ and $n_3 > 1$ in the first and third stages. In case $0 < n_1 < 1$ and $0 < n_3 < 1$ (Fig 5.3F) the first stage reflects decay and the third expresses growth of the density $f_X(x|\Theta)$ given by (5.1) resulting in a "bathtub" shape rather than a trapezoidal shape for the combined density. The graphs in Fig. 5.3 alternate between the three cases $0 < \alpha < 1$ (Fig. 5.3A, D, G and J), $\alpha = 1$ (Figs. 5.3 B, E and H) and

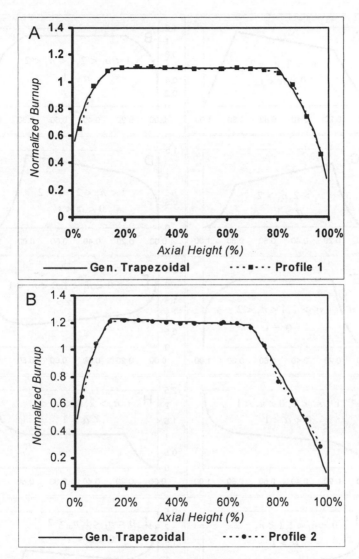

Fig. 5.2 Generalized trapezoidal approximation of axial distributions depicted in Fig. 5.1.
Graph A. $a = 0$, $b = 0.15$, $c = 0.8$, $d = 1$, $n_1 = 1.25$, $n_3 = 1.45$, $\alpha = 1$;
Graph B: $a = 0$, $b = 0.14$, $c = 0.69$, $d = 1$, $n_1 = 1.35$, $n_3 = 1.75$, $\alpha = 1.04$.

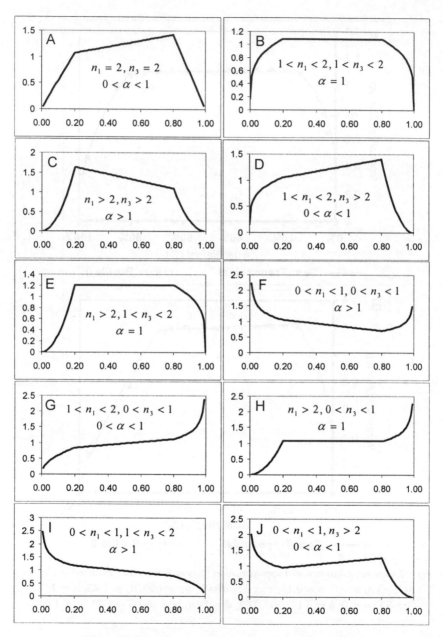

Fig. 5.3 Examples of generalized trapezoidal pdf's (Eq. (5.1)).

$\alpha > 1$ (Figs. 5.3 C, F and I). To appreciate the flexibility due to the parameters n_1 and n_3 only compare Fig. 5.3A with Fig 5.3 J.

5.2.1 Construction of the probability density function

Our approach towards constructing the pdf (5.1) requires to specify: (1) the beginnings and ends of the three stages (a, b, c, d), (2) the growth behavior of the first stage (parameter n_1), (3) the decay behavior of the third stage (parameter n_3) and (4) the relative likelihood of capabilities at the end of the growth stage $[a, b]$ and the beginning of the decay stage $[c, d]$, namely the boundary ratio parameter

$$\alpha = f_X(b)/f_X(c). \tag{5.3}$$

To generalize the trapezoidal distribution we shall take full advantage of the fact that the original trapezoidal pdf (2.22) can be represented as a mixture of three component densities (see, Eqs. (2.25) and (2.26) in Chapter 2). The pdf (5.1) is then naturally constructed using the same mixture technique involving three densities $f_{X_1}, f_{X_2}, f_{X_3}$ with bounded support, such that

$$f_X(x|\Theta) = \sum_{i=1}^{3} \pi_i f_{X_i}(x|\Theta), \quad \sum_{i=1}^{3} \pi_i = 1, \; \pi_i > 0, \tag{5.4}$$

where

$$f_{X_1}(x|\Theta) = f_{X_1}(x|a, b, n_1) = \left(\frac{n_1}{b-a}\right)\left(\frac{x-a}{b-a}\right)^{n_1-1}, \tag{5.5}$$
$$a \leq x < b, \; n_1 > 0,$$

$$f_{X_2}(x|\Theta) = f_{X_2}(x|b, c, \alpha) = \frac{2\{(1-\alpha)x + \alpha c - b\}}{(\alpha+1)(c-b)^2}, \tag{5.6}$$
$$b \leq x \leq c, \alpha > 0,$$

and

$$f_{X_3}(x|\Theta) = f_{X_3}(x|c,d,n_3) = \left(\frac{n_3}{d-c}\right)\left(\frac{d-x}{d-c}\right)^{n_3-1}, \qquad (5.7)$$

$$c \le x < d, \, n_3 > 0.$$

The values of π_i, $i = 1, 2, 3$, are presented below. The pdf's $f_{X_1}(x|a,b,n_1)$ and $f_{X_3}(x|c,d,n_3)$ are chosen for X_1 and X_3 in the aggregated density (5.4) to allow for a nonlinear growth and decay in stages 1 and 3, respectively. The density function $f_{X_2}(x|b,c,\alpha)$ in the second stage is, however, restricted to a linear form such that

$$f_{X_2}(b|b,c,\alpha) = \alpha f_{X_2}(c|b,c,\alpha). \qquad (5.8)$$

For $0 < \alpha < 1 \, (\alpha > 1)$ the density of X_2 in (5.6) exhibits an inclining (declining) behavior. For $\alpha = 1$, it reduces to a uniform density on $[b,c]$. Also note that $f_{X_2}(x|b,c,\alpha)$ and $f_{X_1}(x|a,b,n_1)$ (and $f_{X_3}(x|c,d,n_3)$) share only the boundary parameter b (parameter c). Note that by substituting, $n_1 = n_3 = 2$ and $\alpha = 1$, in the component densities (5.5), (5.6) and (5.7) we obtain successively a linear, a uniform and again a linear form observed in the three consecutive stages of the trapezoidal density provided by Eq (2.26).

The main challenge in the construction of our generalization of the trapezoidal distribution is to select appropriately the remaining mixing probabilities π_1, π_2, π_3 in (5.4) so that the overall density function $f_X(x)$ be continuous. This turns out to be a nontrivial problem.

Theorem 5.1: *The pdf given by (5.1) follows from expressions (5.4), (5.5), (5.6), (5.7) and (5.8) utilizing the mixture probabilities*

$$\begin{cases} \pi_1 = \dfrac{2\alpha(b-a)n_3}{2\alpha(b-a)n_3+(\alpha+1)(c-b)n_1n_3+2(d-c)n_1}, \\[2ex] \pi_2 = \dfrac{(\alpha+1)(c-b)n_1n_3}{2\alpha(b-a)n_3+(\alpha+1)(c-b)n_1n_3+2(d-c)n_1}, \\[2ex] \pi_3 = \dfrac{2(d-c)n_1}{2\alpha(b-a)n_3+(\alpha+1)(c-b)n_1n_3+2(d-c)n_1}, \end{cases} \qquad (5.9)$$

where $a < b < c < d$, $n_1 > 0, n_3 > 0, \alpha > 0$ and, moreover, the pdf given by (5.1) is continuous on the whole interval $[a,d]$.

Proof: Utilizing (5.4), (5.5), (5.6) and (5.7) the density function of the proposed generalized trapezoidal distribution given by (5.1) can be rewritten as

$$
f_X(x|\Theta) = \begin{cases}
\pi_1 f_{X_1}(x|a, b, n_1), & \text{for } a \le x < b \\
\pi_2 f_{X_2}(x|b, c, \alpha), & \text{for } b \le x < c \\
\pi_3 f_{X_3}(x|c, d, n_3), & \text{for } c \le x < d \\
0 & \text{elsewhere,}
\end{cases}
\tag{5.10}
$$

where $\Theta = (a, b, c, d, n_1, n_3, \alpha)$, for some $\pi_i > 0$, $i = 1, 2, 3$ such that $\sum_{i=1}^{3} \pi_i = 1$,

$$
a < b < c < d, \, n_1 > 0, n_3 > 0 \text{ and } \alpha > 0.
\tag{5.11}
$$

It will be convenient to write the three mixture weights π_i, $i = 1, 2, 3$, (which are evidently connected) in the form of a product of two parameters:

$$
\pi_1 = \beta p, \pi_2 = (1 - \beta), \pi_3 = \beta(1 - p),
\tag{5.12}
$$

where $0 < \beta < 1$ and $0 < p < 1$. Hence, $(1 - \beta)$ equals the total probability mass in the central stage of the density (5.10) and p equals the conditional probability of being in the first stage given that one is not in the central stage of the density (5.10). Equation (5.12) assures that

$$
\sum_{i=1}^{3} \pi_i = \beta p + (1 - \beta) + \beta(1 - p) = 1.
\tag{5.13}
$$

From the definition of the boundary ratio parameter α (see Eqs. (5.3) and (5.8)), utilizing (5.5), (5.7), (5.10) and (5.12), we have

$$
\alpha = \frac{f_X^-(b|\Theta)}{f_X^+(c|\Theta)} =
\tag{5.14}
$$

$$
\frac{\beta p f_{X_1}(b|a, b, n_1)}{\beta(1 - p) f_{X_3}(c|c, d, n_3)} = \frac{p(d - c)n_1}{(1 - p)(b - a)n_3},
$$

where $f_X^-(b|\Theta) = \lim_{x \uparrow b} f_X(x|\Theta)$ and $f_X^+(c|\Theta) = \lim_{x \downarrow c} f_X(x|\Theta)$, yielding

$$p = \frac{(b-a)n_3\alpha}{(d-c)n_1 + (b-a)n_3\alpha}. \tag{5.15}$$

Evidently, p does not depend on β. Also the stipulations (5.11) imply $0 < p < 1$.

Continuity of (5.10) at b will follow from the stipulation that

$$f_X^-(b|\Theta) = f_X^+(b|\Theta), \tag{5.16}$$

implying by (5.12) and (5.10) that

$$\beta p f_{X_1}(b|a,b,n_1) = (1-\beta) f_{X_2}(b|b,\,c,\,\alpha). \tag{5.17}$$

Utilizing (5.5), (5.6), (5.17) (namely expressing β as a function of p, $f_{X_1}(\,\cdot\,)$ and $f_{X_2}(\,\cdot\,)$) and (5.15), we arrive at

$$\beta = \frac{2\alpha(b-a)n_3 + 2(d-c)n_1}{2\alpha(b-a)n_3 + (\alpha+1)(c-b)n_1 n_3 + 2(d-c)n_1}. \tag{5.18}$$

From the stipulations in (5.11) it follows that $0 < \beta < 1$. The form of β as given in (5.18) assures continuity of $f_X(\,\cdot\,|\Theta)$ (Eq. (5.10)) at b. Analogously to (5.17) it follows that

$$\beta(1-p) f_{X_3}(c|c,d,n_3) = (1-\beta) f_{X_2}(c|b,\,c,\,\alpha). \tag{5.19}$$

(The reader is advised to check this last derivation.) The continuity of $f_X(\,\cdot\,|\Theta)$ (Eq. (5.10)) at c is implied by (5.12) and (5.19). Substituting (5.15) and (5.18) into (5.12) we arrive, after some straightforward algebraic manipulations, at the mixing probabilities π_i, $i = 1, 2, 3$ as given in (5.9). Finally, substitution of (5.5), (5.6), (5.7) and (5.9) into (5.10), yields (5.1). $\square$

Setting $n_1 = n_3 = 2$ and $\alpha = 1$ in (5.9), we obtain the mixing probabilities given in Eq. (2.27) that accompany the trapezoidal distribution (2.22). Whereas the mixing probabilities in Eq. (2.27) are solely functions of the durations of the three stages of the original trapezoidal density (2.22), in the more general case (5.9), the mixing probabilities are functions of the growth and decay parameters n_1 and n_3 and the boundary ratio parameter α as well. It is illuminating to note that the distance of the first stage $(b-a)$ (third stage $(d-c)$) in (5.9) is weighted by the decay rate n_3 (decay rate n_1) in the third (first) stage.

We note, in passing, that $f_{X_2}(\cdot)$ can be taken to be a conditional TSP (see, Eq. (4.2) in Chapter 4) on $[a, d]$ truncated to $[b, c]$ (rather than the linear form in (5.6)), which would result in a further extension of the trapezoidal distribution presented in Eq. (2.22) permitting oscillation in the central stage. We encourage our readers to pursue the properties of this modification.

5.2.2 Mixing behavior of the component density functions

Some additional insight about the mixing behavior of the component density functions f_{X_i}, $i = 1, 2, 3$, in the generalized trapezoidal pdf (5.10) (or (5.1)) can be gained by studying the limiting behavior of the mixing probabilities presented in (5.9). From (5.18) one easily obtains $\beta = (1 + G)^{-1}$, where

$$G = \frac{(\alpha + 1)(c - b)}{\frac{2(d-c)}{n_3} + \frac{2\alpha(b-a)}{n_1}} . \qquad (5.20)$$

From the conditions on the parameters given in (5.11) we have $G > 0$ and from the relationship between β and G, the largest (least) β corresponds to least (greatest) G.

As $n_1 \to \infty$ and $n_3 \to \infty$, $G \to \infty$ and hence $\beta \downarrow 0$ (the limiting "least" case). Thus, from $\pi_2 = 1 - \beta$ (Eq. (5.12)) it follows that *no* probability mass is attributed to the first and last stages in the limit when $n_1 \to \infty$ and $n_3 \to \infty$ and the density (5.1) converges to the middle part $f_{X_2}(x|b, c, \alpha)$ (Eq. (5.6)).

As $n_1 \downarrow 0$ and $n_3 \downarrow 0$, $G \downarrow 0$ and $\beta \uparrow 1$ (the limiting "greatest" case). Hence, from $\pi_2 = 1 - \beta$ (Eq. (5.12)) it follows that *all* the probability mass is attributed to the first and last stages in the limit as $n_1 \downarrow 0$ and $n_3 \downarrow 0$ (Eqs. (5.5) and (5.7)). It is straightforward to verify that as $n_1 \downarrow 0$ ($n_3 \downarrow 0$) the density $f_{X_1}(x|a, b, n_1)$ (the density $f_{X_3}(x|c, d, n_3)$) converges to a single point mass of 1 at a (at d). Consequently, the density (5.1) converges to a shifted Bernoulli distribution where the probability mass π_1 at a and π_3 at d depend on the relationship between n_1 and n_3 as $n_1 \downarrow 0$ and $n_3 \downarrow 0$. From the mixture probabilities (5.9) it follows that by letting $n_1 \downarrow 0$ and $n_3 \downarrow 0$ while keeping $n_1/n_3 = \kappa$ (constant) we arrive at

$$\begin{cases} \pi_1 \rightarrow & \dfrac{2\alpha(b-a)}{2\alpha(b-a)+2(d-c)\kappa} , \\[4mm] \pi_3 \rightarrow & \dfrac{2(d-c)\kappa}{2\alpha(b-a)+2(d-c)\kappa} . \end{cases} \tag{5.21}$$

Letting $n_1 \downarrow 0$ and keeping n_3 fixed, it follows from (5.9) that in this case $\pi_1 \uparrow 1$, $\pi_2 \downarrow 0$ and $\pi_3 \downarrow 0$. Hence, all the probability mass is attributed to the first stage and the density (5.1) converges to a single points mass of 1 at a. Vice versa, letting $n_1 \rightarrow \infty$ and keeping n_3 fixed, we have $\pi_1 \downarrow 0$. In this case no probability mass is attributed to the first stage and

$$\begin{cases} \pi_2 \rightarrow & \dfrac{(\alpha+1)(c-b)n_3}{(\alpha+1)(c-b)n_3+2(d-c)} , \\[4mm] \pi_3 \rightarrow & \dfrac{2(d-c)}{(\alpha+1)(c-b)n_3+2(d-c)} . \end{cases} \tag{5.22}$$

Consequently, the density (5.1) reduces to a mixture of the two component densities $f_{X_2}(x|\alpha,b,c)$ and $f_{X_3}(x|c,d,n_3)$ assigning the limiting probability π_2 (probability π_3) in (5.22) to the first density (the second density). Figure 5.4 provides some examples of the pdf (5.1) obtained by letting $n_1 \rightarrow \infty$.

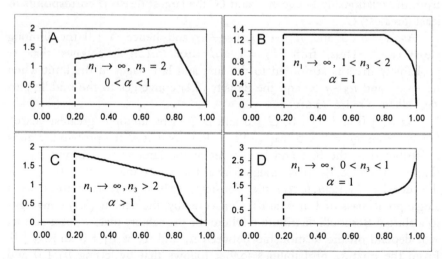

Fig. 5.4 Examples of generalized trapezoidal pdf's (Eq. (5.1))by letting $n_1 \rightarrow \infty$.

With the exception of Fig. 5.3D, the subfigures A, B and C in Figure 5.4 accompany the examples A, B and C in Fig. 5.3. Figure 5.4D is the resulting pdf obtained from Fig. 5.3H when $n_1 \to \infty$. Similar conclusions and plots of the pdfs can be drawn by letting $n_3 \downarrow 0$ and keeping n_1 fixed.

5.3 Basic Properties of the Generalized Trapezoidal Distribution

In the sections below we shall briefly describe the cdf and the moments of the generalized trapezoidal type distributions.

5.3.1 Cumulative distribution function

The derivation of the cdf associated the pdf (5.1) follows most naturally by integration from the structure (5.10), the mixture probabilities (5.9) and the functional form of the component densities (5.5), (5.6) and (5.7), yielding after straightforward algebraic manipulation

$$
F_X(x|\Theta) = \begin{cases}
\frac{\alpha(b-a)\mathcal{C}(\Theta)}{n_1}\left(\frac{x-a}{b-a}\right)^{n_1}, & \text{for } a \le x < b \\[2ex]
\frac{\{\alpha(b-a)+(x-b)n_1\}\mathcal{C}(\Theta)}{n_1} \times, & \text{for } b \le x < c, \\[1ex]
\quad \left\{1+\frac{(\alpha-1)}{2}\frac{(2c-b-x)}{(c-b)}\right\} & \\[2ex]
1-\frac{(d-c)\mathcal{C}(\Theta)}{n_3}\left(\frac{d-x}{d-c}\right)^{n_3}, & \text{for } c \le x < d
\end{cases}
\tag{5.23}
$$

where the normalizing constant $\mathcal{C}(\Theta)$ is given by Eq. (5.2). The reader is strongly encouraged to verify this result in detail. Evidently, the cdf $F_X(x|\Theta)$ vanishes for values of $x < a$ and remains a constant 1 for all $x > d$. Setting $n_1 = n_3 = 2$ and $\alpha = 1$ in (5.23) and (5.2) we obtain the familiar cdf (2.24) presented in Chapter 2 of the trapezoidal distribution.

5.3.2 Moments

Utilizing the component representation (5.4) of the pdf (5.1) and the mixture probabilities $\pi_i, i = 1, 2, 3$ (Eq. (5.9)) we arrive at

$$E[X^k|\Theta] = \pi_1 E[X_1^k|a, b, n_1] + \tag{5.24}$$
$$\pi_2 E[X_2^k|b, c, \alpha] + \pi_3 E[X_3^k|c, d, n_3].$$

The pdf's of the component random variables X_1, X_2 and X_3 are defined in (5.5), (5.6) and (5.7), respectively. From these pdf's we obtain the moments around zero of the component variables:

$$E[X_1^k|a, b, n_1] = \sum_{i=0}^{k} \binom{k}{i} a^{k-i}(b-a)^i \frac{n_1}{n_1 + i}, \tag{5.25}$$

$$E[X_2^k|b, c, \alpha] = \frac{2(1-\alpha)}{(\alpha+1)(c-b)^2} \frac{c^{k+2} - b^{k+2}}{k+2} +$$
$$\frac{2(\alpha c - b)}{(\alpha+1)(c-b)^2} \frac{c^{k+1} - b^{k+1}}{k+1},$$

$$E[X_3^k|c, d, n_3] = \sum_{i=0}^{k} \binom{k}{i} (c-d)^i d^{k-i} \frac{n_3}{n_3 + i}.$$

(Compare with Eq. (2.29) in Chapter 2). Note, the symmetric analogy between $E[X_1^k|a, b, n_1]$ and $E[X_3^k|c, d, n_3]$. Numerical calculations of the k-th moment $E[X^k|\Theta]$ given by (5.24) are quite innocuous employing the modern computer technology and utilizing the closed form expressions for the k-th moment of the component random variables X_1, X_2, X_3 (5.25) and the mixture probabilities $\pi_i, i = 1, 2, 3$ given in (5.9). Deriving closed form expressions for the moments $E[X^k|\Theta]$ of the random variable $X \sim f_X(x|\Theta)$ (Eq. (5.1)) in its general form, although somewhat tedious, is actually straightforward and does not present intrinsic difficulties.

Setting $k = 1$ in (5.25) we obtain for the component variables:

$$E[X_1|a, b, n_1] = \frac{a + n_1 b}{n_1 + 1}, \tag{5.26}$$

$$E[X_2|b, c, \alpha] = \frac{2(1-\alpha)}{(\alpha+1)(c-b)^2} \frac{c^3 - b^3}{3} + \frac{(\alpha c - b)(c + b)}{(\alpha+1)(c-b)}$$

$$E[X_3|c, d, n_3] = \frac{n_3 c + d}{n_3 + 1}.$$

(Compare with Eq. (2.30) in Chapter 2.) Next, the substitution of (5.26) and the mixture probabilities (5.9) into the general expression for the k-th moment (5.24) (using $k = 1$) yields the following expression for the mean of a generalized trapezoidal distribution

$$E[X|\Theta] = C(\Theta) \times \left\{ \frac{\alpha(b-a)(a+n_1 b)}{n_1(n_1+1)} + \right. \tag{5.27}$$
$$\frac{(2+\alpha)c^2 + (\alpha-1)bc - (2\alpha+1)b^2}{6} + \left. \frac{(d-c)(n_3 c + d)}{n_3(n_3+1)} \right\},$$

where the constant $C(\Theta)$ is given by Eq. (5.2) and

$$\Theta = \{a, b, c, d, n_1, n_3, \alpha\}.$$

(Compare with (5.26).) Analogously, we obtain from (5.25) (by substituting $k = 2$ in (5.25)) the second moments around zero of the component variables:

$$E[X_1^2|a, b, n_1] = a^2 + 2a(b-a)\frac{n_1}{n_1+1} + (b-a)^2 \frac{n_1}{n_1+2}, \tag{5.28}$$
$$E[X_2^2|b, c, \alpha] = \frac{2(1-\alpha)}{(\alpha+1)(c-b)^2}\frac{c^4 - b^4}{4} + \frac{2(\alpha c - b)}{(\alpha+1)(c-b)^2}\frac{c^3 - b^3}{3},$$
$$E[X_3^2|c, d, n_3] = d^2 - 2d(d-c)\frac{n_3}{n_3+1} + (d-c)^2 \frac{n_3}{n_3+2}.$$

From here on the expression for second moment around zero of a generalized trapezoidal distribution follows as:

$$E[X^2|\Theta] = C(\Theta) \times \tag{5.29}$$
$$\left[\alpha(b-a)\left\{ \frac{a^2}{n_1} + \frac{2a(b-a)}{n_1+1} + \frac{(b-a)^2}{n_1+2} \right\} + \right.$$
$$\frac{(1-\alpha)}{4}\{c^3 + c^2 b + c\,b^2 + b^3\} + \frac{(\alpha c - b)}{3}\{c^2 + c\,b + b^2\} -$$
$$\left. (c-d)\left\{ \frac{d^2}{n_3} + \frac{2d(c-d)}{n_3+1} + \frac{(c-d)^2}{n_3+2} \right\} \right].$$

Finally, setting $n_1 = n_3 = 2$ and $\alpha = 1$ in (5.27), (5.29) and (5.2) we arrive at the elegant formulas (2.31) and (2.32) in Chapter 2 for the mean and the second moment around zero, respectively, of a trapezoidal distribution with pdf (2.22).

5.4 Concluding Remarks

In the course of the construction of a continuously connected generalized trapezoidal distribution some interesting features have merged, worthy of specific mention. Firstly the structure of these distributions − formally a mixture of three components − differs from the commonly encountered mixtures (see, e.g., Everitt and Hand (1981)) in at least two aspects: (1)- the mixing parameters are of a special form (being a product of two quantities (Eq. (5.12)), each performing a function needed to properly link the three components in Eq. (5.10) in a continuous manner (2)- the components represent different distributions each capable of taking a variety of forms. Next, while classical continuous distributions are usually characterized by the property that continuity is generated by means of a mathematical function that results in a special form of the distribution, in our case continuity is generated by appropriately linking the three relevant parts of the distribution, thus providing an additional flexibility. We have attempted to demonstrate a natural, but so far not well-known, geometrically oriented method of constructing versatile and flexible family of continuous distributions on a compact set. The procedure depends on the values of the parameters of the constituent distributions and illustrates a new form of a mixtures consisting of nonlinear components. The family enjoys transparent physical interpretation and is likely to have potential applications in engineering, communication, behavioral and medical sciences. Estimation of parameter has not been discussed in this chapter. We trust that readers acquainted with Chapters 1-3 will find this an interesting exercise.

Chapter 6

Uneven Two-Sided Power Distributions

Most of the distributions in the previous chapters enjoy certain symmetry characteristics and do not exhibit discontinuities in their densities. The concepts of symmetry and asymmetry permeates a multitude of phenomena in the physical world and play an important role in numerous human activities, in particular in Arts and Sciences. The symmetry of a wheel generates radial symmetry, which is present in many statistical distributions including the basic multivariate Gaussian (or normal) distribution. On the other hand, asymmetric generalizations of the Gaussian family (with a single jump discontinuity) have been available for a long time, perhaps most notably with applications in psychology, communication theory and signal detection (see, e.g., Fechner (1897); Kanefsky and Thomas (1965); Barnard (1989)). It turns out that for the four-parameter TSP distribution introduced in Chapter 4 possessing pdf (4.2) a structure corresponding to an asymmetric Gaussian distribution can be obtained — to be designated as Uneven Two-Sided Power (UTSP) — by appropriately manipulating the central part of generalized trapezoidal distributions with the pdf (5.1) discussed in Chapter 5. The UTSP distribution seems to be suitable for modeling diverse phenomena occurring in financial engineering such as production analysis, standard auction models and equilibrium job search problem. In this chapter its properties are presented and a maximum likelihood (ML) estimation procedures for the threshold location and jump size are developed. A rather elaborated example of an ML procedure is provided utilizing a sample of standardized log differences of bi-monthly US Certificate Deposit interest rates for the period 1966 - 2002. The corresponding time series was constructed using a widely used Auto-Regressive Conditional Heteroscedastic (ARCH) model (see, e.g., Tsay (2002) for a description). The example also seems to demonstrate the practical usefulness and applicability of the concepts and topics developed in the first 5 chapters of this text.

6.1 Motivation

The well-known book by H. Weyl (1952) delineates numerous situations which involve symmetry. In architecture the ancient Greeks were the promoters of symmetry in their classical structures and monuments. In modern arts, the Dutch artist M.C. Escher (1889 - 1972) achieved striking effects in exploring mathematical symmetry (see, e.g., Escher (1989)). The basic symmetry operations: reflection, rotation, double reflection and translation constitute the symmetry group for an object or a figure. It has direct applications in crystallography, amongst other fields. The distinction of symmetries with respect to a given point (center of symmetry), a line (axis of symmetry) or a plane (plane of symmetry) are also important for applications. Human beings and many animals have symmetric proportions. A line from a human's nose to the ground would divide him/her into equal symmetric parts - manifesting bilateral symmetry. The symmetry of a wheel generates radial symmetry, which (as mentioned in the preamble) appears in many statistical distributions including the basic multivariate Gaussian (or normal) distribution. For a more recent, authoritative discussion of the topic of symmetry see Zabell (1988).

Besides the applications of *asymmetric* generalizations of the Gaussian family of distributions (depicted in Fig. 6.1) mentioned above, these types of distributions have also been proposed with an increasing frequency in econometric applications as error terms in linear regression models. Aigner *et al.* (1976) were apparently the first to propose a model with a conditional density jump in the context of production analysis; more recent applications can be found, for example, in standard auction models and the equilibrium job search problems. In standard auction models (see, e.g., Donald and Paarsch (1996)) the density jumps from zero to a positive value and in the equilibrium job search applications it jumps from one level to another, inducing kinks in the cdf (see, e.g., Bowlus *et al.* (2001)). Chernozhukov and Hong (2001) discuss more recently the regression inference problem for the model originally suggested by Aigner *et al.* (1976).

By shrinking the central part of the generalized trapezoidal distribution (pdf (5.1) in Chapter 5 and Fig. 5.4 provide some examples) to a single point Kotz and Van Dorp (2004) arrive at the Uneven Two-Sided Power (UTSP) distribution involving four parameters (in the case when the boundaries, which determine the range, are assumed to be known) with a jump discontinuity at a single point similar to that of the asymmetric generalization of the Gaussian distribution (see, Fig. 6.1). The axis at which the jump discontinuity occurs will be referred to from hereon as the *threshold axis*. The transition from the continuous generalized trapezoidal case to the discontinuous UTSP case can easily be achieved by just one single operation to be demonstrated in the next section.

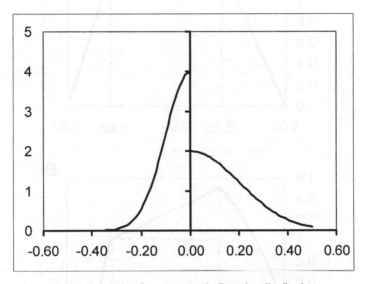

Fig. 6.1 Example of an asymmetric Gaussian distribution.

6.2 Derivation of UTSP Family by a Single Limiting Operation

Recall (Chapter 2) that trapezoidal distributions with pdf (2.22) consisting of three stages are somewhat restrictive, since the growth and decay (in the first and third stages) are limited here to linear functions while the middle stage represents complete (flat) stability rather than a possible (mild) incline or decline. A specific example of a trapezoidal distribution is depicted in Fig. 6.2A.

Generalized trapezoidal distributions inherit the four basic trapezoidal parameters a, b, c and d, and require, for its complete description, two additional parameters n_1 and n_3 specifying the (not necessarily linear) growth and decay rates at the first and third stages of the distribution

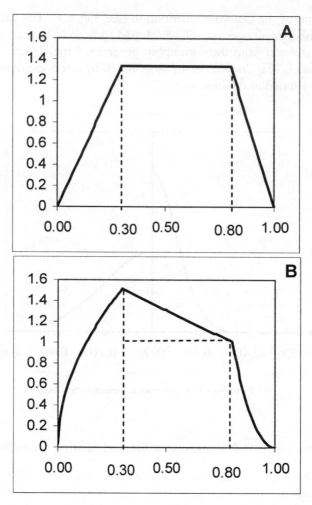

Fig. 6.2 A: A trapezoidal distribution with parameters $a = 0, b = 0.3, c = 0.8$ and $d = 1$;
B: A generalized trapezoidal distribution with parameters $a = 0, b = 0.3, c = 0.8, d = 1$,
$n_1 = 1.5, n_3 = 3$, and $\alpha = 1.5$.

respectively, and also the boundary ratio parameter $\alpha > 0$ satisfying Eq. (5.3) in Chapter 5. As shown in Chapter 5, Sec. 5.2 the density function of a generalized trapezoidal distribution is given by (5.1), where $a < b < c < d$, $n_1 > 0, n_3 > 0$ and $\alpha > 0$. (Here the growth and decay may exhibit a nonlinear convex or concave behavior and $f_X(b)$ and $f_X(c)$ do not

166

necessarily take the same value — recall the boundary ratio parameter $\alpha = f_X(b)/f_X(c)$ given in Eq. (5.3).) Figure 6.2B provides a graph of the pdf of a standard generalized trapezoidal distribution with support $[0, 1]$. The three stages of both trapezoidal distributions and their generalization described by the pdf (5.1) are indicated in Fig. 6.2 by vertical dotted lines.

Figure 6.3 displays the resulting UTSP distribution generated from the generalized trapezoidal distribution in Fig 6.2B by collapsing the central part in the pdf (5.1) to a single point. The functional form of a UTSP pdf, obtained by letting $c \downarrow b$ in the pdf (5.1) is

$$f_X(x|a, b, d, n_1, n_3, \alpha) = \qquad\qquad (6.1)$$

$$\begin{cases} \frac{\alpha n_1 n_3}{\alpha(b-a)n_3+(d-b)n_1}\left(\frac{x-a}{b-a}\right)^{n_1-1}, & \text{for } a \leq x < b \\ \frac{n_1 n_3}{\alpha(b-a)n_3+(d-b)n_1}\left(\frac{d-x}{d-b}\right)^{n_3-1}, & \text{for } b \leq x < d \\ 0, & \text{elsewhere,} \end{cases}$$

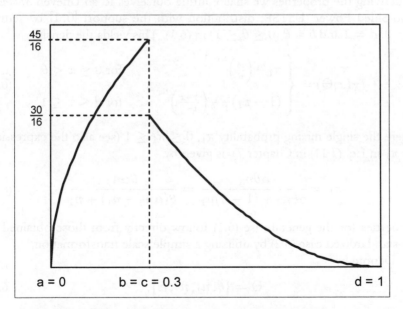

Fig. 6.3 Example of an Uneven Two-Sided Power (UTSP) distribution with parameters $a = 0, b = 0.3, \ d = 1, n_1 = 1.5, \ n_3 = 3$ and $\alpha = 1.5$.

with support $[a, d]$ and the corresponding cdf is given by

$$F_X(x|a, b, d, n_1, n_3, \alpha) =$$
$$\begin{cases} 0, & \text{for } x < a \\ \frac{\alpha(b-a)n_3}{\alpha(b-a)n_3+(d-b)n_1} \left(\frac{x-a}{b-a}\right)^{n_1}, & \text{for } a \leq x < b \\ 1 - \frac{(d-b)n_1}{\alpha(b-a)n_3+(d-b)n_1} \left(\frac{d-x}{d-b}\right)^{n_3}, & \text{for } b \leq x < d \\ 1, & \text{for } x > d. \end{cases}$$

Substituting $\alpha = 1$ and $n_1 = n_3 = n$ into (6.1) we arrive at a four-parameter TSP distribution with the pdf (4.2) discussed in Chapter 4. Recall that a four-parameter TSP distribution was in turn obtained by generalizing the triangular family of distributions (1.1) presented in Chapter 1.

6.3 Some Properties of the Uneven STSP Distribution

In deriving the properties we shall confine ourselves to an Uneven *Standard* Two-Sided Power (USTSP) distribution with the support $[0, 1]$ by setting $a = 0, d = 1$ and $b = \theta$ $(0 \leq \theta \leq 1)$ in (6.1). This yields the density

$$f_X(x|\Theta) = \begin{cases} \pi_1 \frac{n_1}{\theta} \left(\frac{x}{\theta}\right)^{n_1-1}, & \text{for } 0 \leq x < \theta \\ (1 - \pi_1)\frac{n_3}{1-\theta} \left(\frac{1-x}{1-\theta}\right)^{n_3-1}, & \text{for } \theta \leq x \leq 1, \end{cases} \tag{6.2}$$

where the single mixing probability π_1, $0 \leq \pi_1 \leq 1$ (see also the expression for π_1 in Eq. (7.11) in Chapter 7) is given by

$$\pi_1 = \frac{\alpha\theta n_3}{\alpha\theta n_3 + (1 - \theta)n_1} = \frac{\theta\alpha n_3}{\theta(\alpha n_3 - n_1) + n_1}. \tag{6.3}$$

Properties for the general case (6.1) follow directly from those obtained in the standardized case (6.2) by utilizing a simple scale transformation.

Denote by

$$\Theta = \{\theta, n_1, n_3, \alpha\}, \tag{6.4}$$

a vector of the four parameters, where $0 \leq \theta \leq 1$, $n_1, n_3 > 0$ and $\alpha > 0$. The possible geometrical shapes of the USTSP pdf given by (6.2) are similar to those of the STSP distribution (3.11) (see Fig. 3.4 in Chapter 3) including the J-shaped and U-shaped forms while, in addition, allowing here for a jump discontinuity at the threshold θ. Note that it follows from (6.2) and (6.3) that, for example, as $n_3 \to \infty$ the mixing probability $\pi_1 \to 1$. This is because for all x such that $\theta \le x \le 1$, $(1 - x)/(1 - \theta) \le 1$ in the second branch of (6.2). The limiting behavior of $f_X(x|\Theta)$ and the mixture probability π_1 as a function of its parameters will be discussed in more detail at the end of this section.

Setting $\alpha = 1$ in (6.2) yields a continuous generalization of the STSP pdf (3.15) (with two parameters) in Chapter 3 with the pdf:

$$f_X(x|\theta, n_1, n_3) = \begin{cases} \frac{n_1 n_3}{\theta n_3 + (1-\theta) n_1} \left(\frac{x}{\theta}\right)^{n_1 - 1}, & \text{for } 0 \le x < \theta \\ \frac{n_1 n_3}{\theta n_3 + (1-\theta) n_1} \left(\frac{1-x}{1-\theta}\right)^{n_3 - 1}, & \text{for } \theta \le x \le 1, \end{cases} \tag{6.5}$$

allowing for different powers n_1 and n_3 in the two respective branches of the STSP density (3.11). We shall refer to the three-parameter densities given by (6.5) as Generalized STSP (GSTSP) distributions. In the limiting cases $\theta = 1$ and $\theta = 0$, the pdf (3.11) or the pdf (6.5) simplify to a power distribution or its reflection, respectively. The reader is advised to study Fig. A.1 and Table A.1 in Appendix A to further clarify and analyze the relationships between the Triangular, Trapezoidal, TSP, GTSP, UTSP and Generalized Trapezoidal distributions. The five-parameter GTSP distribution may be obtained from (6.5) using a linear scale transformation $(b - a)X + a$ and appears in a reparameterized form in Schmeiser and Lal (1985).

The cdf associated with density (6.2) is continuous but non-differentiable at the threshold axis at θ and is given by

$$F_X(x|\Theta) = \begin{cases} 0, & \text{for } x \le 0 \\ \pi_1 \left(\frac{x}{\theta}\right)^{n_1}, & \text{for } 0 < x < \theta \\ 1 - (1 - \pi_1)\left(\frac{1-x}{1-\theta}\right)^{n_3}, & \text{for } \theta \le x < 1 \\ 1, & \text{for } x \ge 1. \end{cases} \tag{6.6}$$

From (6.6) (the cdf of the USTSP distribution) we obtain that

$$F_X(\theta|\Theta) = \pi_1. \tag{6.7}$$

Hence, the total probability mass is split into two parts π_1 and $(1 - \pi_1)$ at the threshold θ (Eq (6.3)). Recall that for the STSP distribution (Eq. (3.11) in Chapter 3) expression (6.7) simplifies to $F_X(\theta|\Theta) = \theta$ regardless of the value of n. The latter property is referred to as the "hinge" property of the STSP family (see Property 2 in Sec. 3.2.3).

From (6.2) and (6.3) we have

$$\frac{f_X^-(\theta|\Theta)}{f_X^+(\theta|\Theta)} = \frac{\lim\limits_{x \uparrow \theta} f_X(x|\Theta)}{\lim\limits_{x \downarrow \theta} f_X(x|\Theta)} = \frac{\pi_1 \frac{n_1}{\theta}}{(1-\pi_1)\frac{n_3}{1-\theta}} = \alpha.$$

While in case of a generalized trapezoidal distribution α was referred to as a boundary ratio parameter, for the USTSP distribution α could be interpreted as a *jump* parameter. In case $\alpha = 1$, there is no jump at the threshold θ, in case $\alpha > 1$ ($\alpha < 1$) the density jumps down (up) at θ, with larger (smaller) values indicating a larger (smaller) jump down (up) of the density. The size of the jump discontinuity at the threshold θ may be directly derived utilizing (6.2) and the definition of the mixing probability π_1 (6.3) to be

$$|f_X^-(\theta|\Theta) - f_X^+(\theta|\Theta)| = \frac{|1-\alpha|\alpha n_1 n_3}{\alpha \theta n_3 + (1-\theta)n_1}.$$

(Observe that for USTSP distributions $a = 0, d = 1$ and $b = \theta$.) The reader is advised to develop graphs of USTSP distributions for a) $0 < \alpha < 1$, b) $\alpha = 1$ and c) $\alpha > 1$.

The k-th moment of USTSP distributions (6.2) follow directly using the (inherited) mixture structure (5.4) in Chapter 5 of the generalized trapezoidal distribution (*without the central component X_2*), the mixing weight π_1 (6.3), and the k-th moment of a (one-sided) power distribution on $[0, \theta]$

$$\frac{n_1 \theta^k}{n_1 + k} \tag{6.8}$$

as well as the k-th moment of the reflected power distribution on $[\theta, 1]$

$$n_3 \sum_{i=0}^{k} \binom{k}{i} \frac{(\theta-1)^i}{n_3 + i}, \tag{6.9}$$

yielding

$$E[X^k|\Theta] = \pi_1 \frac{n_1 \theta^k}{n_1 + k} + \tag{6.10}$$

$$(1 - \pi_1) \left[n_3 \sum_{i=0}^{k} \binom{k}{i} \frac{(\theta - 1)^i}{n_3 + i} \right], k = 1, 2, \ldots .$$

Consequently, (substituting $k = 1$)

$$E[X|\Theta] = \pi_1 \left[\frac{n_1 \theta}{n_1 + 1} \right] + (1 - \pi_1) \left[\frac{n_3 \theta + 1}{n_3 + 1} \right]. \tag{6.11}$$

The mean value of a power distribution on $[0, \theta]$ (of a reflected power distribution on $[1, \theta]$) in (6.11) also follows directly from the mean value formula (4.6) of a TSP distribution in Chapter 4 by substituting $a = 0$, $m = \theta, b = \theta$ (substituting $a = \theta, m = \theta, b = 1$) in (6.7). Observe the substantial difference in the structure of Eq. (6.8) as compared to Eq. (6.9). This emphasizes the intrinsic differences between a power distribution on $[0, \theta]$ and a reflected power distribution on $[\theta, 1]$.

The second moment is obtained from (6.10) by setting $k = 2$, yielding

$$E[X^2|\Theta] = \pi_1 \left[\frac{n_1(n_1 + 1)\theta^2}{(n_1 + 2)(n_1 + 1)} \right] + \tag{6.12}$$

$$(1 - \pi_1) \left[\frac{2 + 2n_3\theta + n_3(n_3 + 1)\theta^2}{(n_3 + 2)(n_3 + 1)} \right].$$

Substituting for π_1 as given by (6.3) into (6.11) and (6.12) we arrive at the expressions for the first two moments in terms of the four parameters n_1, n_3, θ and α, respectively. A derivation of a closed form expression for the variance (σ^2) is somewhat tedious, but straightforwardly follows utilizing (6.3), (6.11) and (6.12) and the definition

$$\sigma^2 = E[X^2|\Theta] - E^2[X|\Theta]. \tag{6.13}$$

Using modern computational facilities one may easily calculate higher moments of USTSP distribution from (6.10) including skewness and kurtosis, which are of practical importance.

The behavior of the mixing probability π_1 given by (6.3) as a function of the jump parameter α and of the threshold parameter θ — at which the jump occurs — is of special interest. Figure 6.4 displays the mixture probability π_1 as function of θ for five different values of α from 0.1 up to 10 for the case when $n_1 = n_3 = 2$. Compare the case $\alpha = 10$ in Fig. 6.4 corresponding to a small value of π_1 for a bulk of the values of θ with the case $\alpha = 0.1$ in which case $\pi_1 \geq 0.8$ for all $\theta \in [0.29, 1]$.

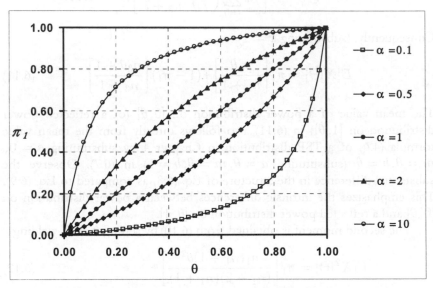

Fig. 6.4 Behavior of mixture probability π_1 given by (6.3) as a function of the threshold parameter θ for different values of the jump parameter α, with $n_1 = n_3 = 2$.

Setting $\alpha = 1$, $n_1 = n_3 = 2$ in (6.2) and (6.3) results in the triangular distribution − in this case $\pi_1 = \theta$ − with the property that the probability mass to the left of the mode equals the distance of the mode to the lower bound relative to the total range of the support. This property is preserved by the TSP generalization (4.2) (in Chapter 4) of the triangular distribution and follows from (6.7) by substituting $\alpha = 1$, $n_1 = n_3 = n$ into (6.3). Hence from Fig. 6.4 we conclude that in the case of the USTSP distribution with $n_1 = n_3 = n$ (or equivalently, $n_1/n_3 = 1$) the probability mass to the left of the threshold parameter θ is less (larger) than its relative distance from the lower bound when the density jumps up, i.e. $\alpha < 1$, (down, i.e. $\alpha > 1$). Finally, when $(n_1/n_3) < 1$ (> 1) in an USTSP distribution (Eqs. (6.2) and (6.3)) smaller (larger) probability mass is assigned to the left of the threshold parameter θ than it is in the case $n_1/n_3 = 1$. The limiting behavior of the mixing probability π_1 as a function of one of the parameters n_1, n_3, α and θ, while keeping the others fixed, follows directly from (6.3). In fact, $\pi_1 \downarrow 0$ ($\pi_1 \uparrow 1$) when $n_1 \to \infty$, or $n_3 \downarrow 0$, or $\alpha \downarrow 0$, or $\theta \downarrow 0$ (when $n_1 \downarrow 0$, or $n_3 \to \infty$, or $\alpha \to \infty$, or $\theta \uparrow 1$). Note the case $\alpha = 1$ in Fig. 6.4.

The behavior of the mean $E[X|\Theta]$ (6.11) as a function of the jump parameter α (while keeping the other parameters fixed) follows directly from (6.11) and behavior of the mixing weight π_1(6.3). In fact, when α increases, the mixing weight π_1 increases, assigning a larger weight to the mean value of the power distribution on $[0, \theta]$ in (6.11) (and a smaller weight to the mean value of the reflected power distribution on $[\theta, 1]$) and hence results in a decrease of $E[X|\Theta]$. In a similar manner one can derive the behavior of $E[X|\Theta]$ as a function of the other parameters.

Table 6.1 summarizes the limiting behavior of the mixture probability π_1 (6.3), the limiting pdf of (6.2) with the mean $E[X|\Theta]$ (6.11) for various limiting scenarios. All the parameter components in the parameter vector Θ (6.4) that <u>do not</u> appear in the first column of Table 6.1 are assumed to be fixed. Some brief comments on the informative results presented in Table 6.1 are in order. The scenarios 1) $\theta \downarrow 0$; 2) $n_1 \to \infty$; 3) $\theta \uparrow 1$ and 4) $n_3 \to \infty$ all result in a single point limiting density at the values of θ (specifically, limiting values 0 or 1 of θ in the cases 1 and 3, respectively). Also the last two limiting scenarios in Table 6.1 (the 9-th and 10-th row) keeping $n_1/n_3 = \beta$ constant result in the same value of the mixture probability π_1 (since π_1 as given in (6.3) depends on the ratio n_1/n_3) but yield very different limiting distributions: a single point mass at θ when $n_1, n_3 \to \infty$ and a two-point Bernoulli distribution (with parameter π_1) at 0 and 1 when $n_1, n_3 \downarrow 0$. This is because when $n_1, n_3 \downarrow 0$ the structure of the original pdf (6.2) becomes U-shaped with an antimode at θ. The two situations $\alpha \to \infty$ and $n_3 \downarrow 0$ result in the same limiting density $(x/\theta)^{n_1}$ on $[0, \theta]$ and equivalently the situations $\alpha \downarrow 0$ and $n_1 \downarrow 0$ both yield the density $\{(1 - x)/(1 - \theta)\}^{n_3-1}$ on $[\theta, 1]$. The reader is encouraged to devise a graphical representation of Table 6.1.

6.4 ML Estimation Procedure for USTSP Distributions

In this section we shall derive a maximum likelihood procedure for USTSP distributions that is algorithmically straightforward in terms of elementary function's evaluations. The reader is advised to compare the discussion below with similar discussions of ML procedures dealing with other classes of TSP distributions in Chapters 3 and 4.

Table 6.1. Limiting behavior of the mixture probability π_1 (6.3),
the pdf (6.2) and the mean $E[X|\Theta]$ (6.11) under a variety of scenarios.
Parameters not mentioned in the first column are assumed to be fixed.

Scenario	π_1	$f_X(x|\theta, n_1, n_3)$	$E[X|\Theta]$
$\alpha \to \infty$	$\uparrow 1$	$\to \left(\frac{x}{\theta}\right)^{n_1-1}$ on $[0,\theta]$	$\downarrow \frac{n_1\theta}{n_1+1}$
$\theta \downarrow 0$	$\uparrow 1$	single point mass at 0	$\downarrow 0$
$n_1 \to \infty$	$\uparrow 1$	single point mass at θ	$\downarrow \theta$
$n_3 \downarrow 0$	$\uparrow 1$	$\to \left(\frac{x}{\theta}\right)^{n_1-1}$ on $[0,\theta]$	$\downarrow \frac{n_1\theta}{n_1+1}$
$\alpha \downarrow 0$	$\downarrow 0$	$\to \left(\frac{1-x}{1-\theta}\right)^{n_3-1}$ on $[\theta,1]$	$\uparrow \frac{n_3\theta+1}{n_3+1}$
$\theta \uparrow 1$	$\downarrow 0$	single point mass at 1	$\uparrow 1$
$n_1 \downarrow 0$	$\downarrow 0$	$\to \left(\frac{1-x}{1-\theta}\right)^{n_3-1}$ on $[\theta,1]$	$\uparrow \frac{n_3\theta+1}{n_3+1}$
$n_3 \to \infty$	$\downarrow 0$	single point mass at θ	$\uparrow \theta$
$n_1, n_3 \to \infty,$ $\frac{n_1}{n_3} = \beta$	$= \frac{\alpha\theta}{\alpha\theta+\beta(1-\theta)}$	single point mass at θ	$\to \theta$
$n_1, n_3 \downarrow 0$ $\frac{n_1}{n_3} = \beta$	$= \frac{\alpha\theta}{\alpha\theta+\beta(1-\theta)}$	Bernoulli $(\pi_1, 1-\pi_1)$ at 0 and 1	$\to 1 - \pi_1$

Let for a random sample of size m with the values $\underline{X} = (X_1, \ldots, X_m)$ the order statistics be $X_{(1)} < X_{(2)} < \ldots < X_{(m)}$. Utilizing the formulas for the density of USTSP distributions (6.2) and (6.3), the likelihood $\mathcal{L}(\underline{X}|\Theta)$ for $\underline{X}$ is, by definition,

$$\prod_{i=1}^{r} \frac{\alpha n_1 n_3}{\alpha \theta n_3 + (1-\theta) n_1} \left(\frac{X_{(i)}}{\theta} \right)^{n_1 - 1} \times \qquad (6.14)$$

$$\prod_{i=r+1}^{m} \frac{n_1 n_3}{\alpha \theta n_3 + (1-\theta) n_1} \left(\frac{1 - X_{(i)}}{1 - \theta} \right)^{n_3 - 1},$$

where the integer r is defined so that $X_{(r)} \leq \theta < X_{(r+1)}$, with $X_{(0)} \equiv 0$, $X_{(m+1)} \equiv 1$. Collecting the terms in (6.14) we obtain

$$\mathcal{L}(\underline{X} \mid \Theta) = (n_1 n_3)^m \left[\frac{\alpha^r}{\{\alpha \theta n_3 + (1-\theta) n_1\}^m} \right] \times \qquad (6.15)$$

$$\left[\prod_{i=1}^{r} \frac{X_{(i)}}{\theta} \right]^{n_1 - 1} \left[\prod_{i=r+1}^{m} \frac{1 - X_{(i)}}{1 - \theta} \right]^{n_3 - 1}.$$

The main difficulty in maximizing (6.15) as a function of parameters n_1, n_3, α and θ is due to an irregular behavior of $\mathcal{L}(\underline{X} \mid \Theta)$ as a function of the threshold parameter θ. Figure 1.5 in Chapter 1 depicts an example of $\mathcal{L}(\underline{X} \mid \Theta)$ as function of the θ for the case $n_1 = n_3 = 2$ and $\alpha = 1$ and eight ($m = 8$) order statistics given in Eq. (1.38) in Chapter 1 (containing multiple local maxima). When $n_1 = n_3 = n$ and $\alpha = 1$ in (6.15), $\mathcal{L}(\underline{X} \mid \Theta)$ reduces to the likelihood associated with the STSP distribution given by Eq. (3.11). In Chapter 3 (Theorem 5.2, Sec. 5.3) we have shown for the STSP distribution that under the condition $n \geq 1$, the maximum of $\mathcal{L}(\underline{X} \mid \Theta)$ as a function of θ is attained at one of the order statistics $X_{(i)}$, $i = 1, \ldots, m$, (as indicated in Fig. 5.5). In fact, maximization of $\mathcal{L}(\underline{X} \mid \Theta)$ for the case that one can *a priori* assess $n_1 \geq 1$ and $n_3 \geq 1$ can be achieved by devising an appropriate numerical algorithm which results in the ML estimators $\widehat{\alpha}$, $\widehat{n_1}$, $\widehat{n_3}$ and $\widehat{\theta}$. A unimodal histogram of the data under consideration could confirm the precondition $n_1 \geq 1$ and $n_3 \geq 1$ of the numerical algorithm. The procedure presented below seems to be quite intuitive and direct:

Iteration $k : (k = 1, 2, \ldots)$
(Don't confuse this ordinal k with the order of the moments in (6.10)!)
Step 1: Given $(n_3)_k, \alpha_k$, and θ_k, maximize $\mathcal{L}(\underline{X} \mid \Theta)$ for $(n_1)_{k+1}$
Step 2: Given $(n_1)_{k+1}, \alpha_k$, and θ_k, maximize $\mathcal{L}(\underline{X} \mid \Theta)$ for $(n_3)_{k+1}$
Step 3: Given $(n_1)_{k+1}$ and $(n_3)_{k+1}$ and θ_k, maximize $\mathcal{L}(\underline{X} \mid \Theta)$ for α_{k+1}

Step 4: Given $(n_1)_{k+1}$ and $(n_3)_{k+1}$ and α_{k+1}, maximize $\mathcal{L}(\underline{X} \mid \Theta)$ for θ_{k+1}
Step 5: Go back to *Step* 1, unless the pre-assigned convergence criterion
       has already been met.

Here $(n_1)_k$, $(n_3)_k$ denote the values n_1 and n_3 at the k-th iteration, respectively. The convergence criterion in Step 5 corresponds to a failure of an increase in $\mathcal{L}(\underline{X} \mid \Theta)$ at a pre-assigned tolerance level. A natural starting point for the algorithm to maximizing $\mathcal{L}(\underline{X} \mid \Theta)$ are the ML estimators for the STSP distribution (3.11) (for the case $n \geq 1$) given in Chapter 3, i.e.

$$\begin{cases} \theta_0 = X_{(\widehat{r})} \\ (n_1)_0 = (n_3)_0 = -\dfrac{s}{Log\, M(\widehat{r})} \\ \alpha_0 = 1 \end{cases} \tag{6.16}$$

where $\widehat{r}$ and $M(r)$ are given by Eqs. (3.29) and (3.30) in Chapter 3, respectively. Among the first four steps in the k-th iteration described above, Step 4 is the most cumbersome (although straightforward) since it requires maximization of the likelihood (6.15) over $m + 1$ disjoint intervals $X_{(r)} \leq \theta < X_{(r+1)}$, $r = 0, \ldots, m$, with $X_{(0)} \equiv 0$, $X_{(m+1)} \equiv 1$. Details about these four steps are presented in the next subsection. It is important to note here that when the algorithm above converges to a solution where either $\widehat{n}_1$ or $\widehat{n}_3$ is less than 1 (i.e. inconsistent with the precondition $n_1 \geq 1$ and $n_3 \geq 1$) one cannot interpret the converged estimates $\widehat{\alpha}$, $\widehat{n}_1$, $\widehat{n}_3$ and $\widehat{\theta}$ as the *ML estimators* for the data under consideration.

    At a first read the reader may wish to skip the next subsection and proceed to Sec. 6.5 which presents an illustrative example of the use of the ML procedure above utilizing the monthly USA certificate deposit rates for the period 1966-2002. Finally note that the numerical algorithm described above can easily be modified (by omitting Step 3) to provide a maximum likelihood estimation procedure for a GSTSP distribution given by the density (6.5) which does not involve α.

6.4.1 Mathematical details of the ML estimation procedure

Some details related to the first four steps in the k-th iteration presented in Sec. 6.4 and the ML procedure maximizing the likelihood given by (6.15) are provided below. (Compare this description with the procedure described by Theorem 3.2 in Sec. 3.3 for a simpler case and note the modifications that have been incorporated.)

6.4.1.1 Step 1: maximizing over the LHS power parameter n_1

Two cases 1.A : $X_1 \leq \ldots \leq X_{(r)} \leq \theta < X_{(r+1)}$ and 1.B : $\theta < X_{(1)}$ ought to be considered since these cases represent two different forms of the corresponding likelihood function.

Case 1.A : Assuming $X_1 \leq \ldots \leq X_{(r)} \leq \theta < X_{(r+1)}$ and introducing the notation

$$0 < \mathcal{A} = \prod_{i=1}^{r} \frac{X_{(i)}}{\theta} < 1, \quad \mathcal{B} = (1 - \theta) > 0 \text{ and } \mathcal{C} = \alpha\theta n_3 > 0, \quad (6.17)$$

we may rewrite the likelihood function (6.15) in the more compact form

$$\mathcal{L}(\underline{X} \mid n_1) \propto \mathcal{A}^{n_1 - 1} \left\{ \frac{n_1}{\mathcal{C} + \mathcal{B}n_1} \right\}^m. \quad (6.18)$$

From the conditions (6.17) and the expression (6.18) it follows that

$$\begin{cases} \mathcal{L}(\underline{X} \mid n_1) = 0, & \text{for } n_1 = 0 \\ \mathcal{L}(\underline{X} \mid n_1) > 0, & \text{for } n_1 > 0 \\ \mathcal{L}(\underline{X} \mid n_1) \to 0, & \text{as } n_1 \to \infty. \end{cases} \quad (6.19)$$

Hence $\mathcal{L}(\underline{X} \mid n_1)$ (6.18) attains its maximum at some stationary point $n_1^* > 0$. Instead of maximizing $\mathcal{L}(\underline{X} \mid n_1)$ we equivalently maximize its logarithm

$$(n_1 - 1)Log(\mathcal{A}) + mLog(n_1) - mLog(\mathcal{C} + \mathcal{B}n_1). \quad (6.20)$$

Setting $dLog(\mathcal{L}(X \mid n_1))/dn_1 = 0$ leads to the equation:

$$\mathcal{B}Log(\mathcal{A})(n_1)^2 + \mathcal{C}Log(\mathcal{A})n_1 + m\mathcal{C} = 0. \quad (6.21)$$

Since the quadratic equation (6.21) in n_1 possesses at most two real-valued solutions, we conclude, utilizing (6.19), that $\mathcal{L}(\underline{X} \mid n_1)$ has a unique stationary point $n_1^* > 0$, which may explicitly be obtained from (6.21) substituting the designations (6.17).

Case 1.B : Assuming $\theta < X_{(1)}$ and rewriting (6.15) using the designations (6.17) we have

$$\mathcal{L}(\underline{X} \mid n_1) \propto \left\{ \frac{n_1}{\mathcal{C} + \mathcal{B}n_1} \right\}^m. \tag{6.22}$$

Note that, the term $\mathcal{A}^{n_1-1}$ in (6.18) is not included in (6.22). From (6.22) it follows, taking (6.17) into account, that

$$\begin{cases} \mathcal{L}(\underline{X} \mid n_1) = 0, & \text{for } n_1 = 0 \\ \mathcal{L}(\underline{X} \mid n_1) > 0, & \text{for } n_1 > 0 \\ \mathcal{L}(\underline{X} \mid n_1) \to \mathcal{B}^{-m} > 0, & \text{as } n_1 \to \infty. \end{cases} \tag{6.23}$$

Moreover,

$$\frac{d\mathcal{L}(\underline{X} \mid n_1))}{dn_1} = m \left\{ \frac{n_1}{\mathcal{C} + \mathcal{B}n_1} \right\}^{m-1} \frac{\mathcal{C}}{(\mathcal{C} + \mathcal{B}n_1)^2} > 0 \tag{6.24}$$

for $n_1 > 0$, since $\mathcal{C}$ and $\mathcal{B}$ are both positive quantities. Hence, Case 1.B does not yield a maximum solution for the likelihood $\mathcal{L}(X \mid n_1)$. As above, one could actually rule this situation out by using the starting solution as given by (6.16) (corresponding to the STSP case) or a histogram of the data of an unimodal form which contains observations on both sides of the mode. If however this case does occur, no solution can be obtained and the algorithm terminates.

6.4.1.2 Step 2: maximizing over the RHS power parameter n_3

As before, we shall separately consider the two cases yielding two distinct forms of the likelihood 2.A : $X_{(r)} \le \theta < X_{(r+1)} < \ldots < X_{(m)}$ and 2.B : $\theta \ge X_{(m)}$.

Case 2.A: Assuming $X_{(r)} \le \theta < X_{(r+1)} < \ldots < X_{(m)}$ and introducing the notation

$$0 < \mathcal{D} = \prod_{i=r+1}^{m} \frac{1 - X_{(i)}}{1 - \theta} < 1, \; \mathcal{E} = \alpha\theta > 0 \text{ and } \mathcal{F} = (1 - \theta)n_1 > 0$$

(compare with (6.17)) we rewrite the likelihood in (6.15) as

$$\mathcal{L}(\underline{X} \mid n_3) \propto \mathcal{D}^{n_3-1} \left\{ \frac{n_3}{\mathcal{F} + \mathcal{E}n_3} \right\}^m.$$

Analogously to (6.18) dealing with the LHS power parameter n_1, by taking the derivative of $Log\{\mathcal{L}(\underline{X} \mid n_3)\}$, we obtain a unique solution $n_3^* > 0$, maximizing $\mathcal{L}(\underline{X} \mid n_3)$, by solving the quadratic equation

$$\mathcal{E}Ln(\mathcal{D})(n_3)^2 + \mathcal{F}Ln(\mathcal{D})n_3 + m\mathcal{F} = 0.$$

As in Step 1 (Sec. 6.4.1.1) no solution can be obtained here for Case 2.B and the algorithm terminates.

6.4.1.3 Step 3: maximizing over the jump parameter α

Introducing the notation

$$\mathcal{G} = \theta n_3 > 0 \text{ and } \mathcal{H} = (1 - \theta)n_1 > 0 \qquad (6.25)$$

we now rewrite the likelihood (6.15) as

$$\mathcal{L}(\underline{X} \mid \alpha) \propto \frac{\alpha^r}{\{\mathcal{G}\alpha + \mathcal{H}\}^m}. \qquad (6.26)$$

As above, instead of maximizing $\mathcal{L}(\underline{X} \mid \alpha)$, we can equivalently maximize its logarithm

$$rLog(\alpha) - mLog\{\mathcal{G}\alpha + \mathcal{H}\}.$$

Setting $dLog\{\mathcal{L}(\underline{X} \mid \alpha)\}/d\alpha = 0$, we arrive at

$$(r - m)\mathcal{G}\alpha + r\mathcal{H} = 0. \qquad (6.27)$$

From $(r - m) < 0$ (since the maximum over θ is a *priori* assessed to be attained between $X_{(1)}$ and $X_{(m)}$ due to a unimodal histogram of the data), and

$$Sign[dLog\{\mathcal{L}(\underline{X} \mid \alpha)\}/d\alpha] = Sign\{(r - m)\mathcal{G}\alpha + r\mathcal{H}\},$$

it follows from (6.27) and taking notation (6.25) into account that

$$\alpha^* = \frac{r\mathcal{H}}{(m-r)\mathcal{G}} = \frac{r\theta n_3}{(m-r)(1-\theta)n_1} > 0$$

maximizes $\mathcal{L}(\underline{X} \mid \alpha)$ in (6.26). (Note that an increase in θ increases the value of the jump parameter α as indicated in Sec. 6.3.)

6.4.1.4 Step 4: maximizing over the threshold parameter θ

Here we shall minimize with respect to θ the reciprocal of $\mathcal{L}(X \mid \theta)$ given by (6.15)

$$\frac{1}{\mathcal{L}(\underline{X} \mid \theta)} \propto \{(\alpha n_3 - n_1)\theta + n_1\}^m \, \theta^{(n_1-1)r}(1-\theta)^{(n_3-1)(m-r)} \qquad (6.28)$$

over the set

$$X_{(r)} \leq \theta < X_{(r+1)}, 0 \leq r \leq m, \qquad (6.29)$$

with $X_{(0)} \equiv 0$, $X_{(m+1)} \equiv 1$. The difficulty in minimizing (6.28) over the set of values (6.29) is that $(m+1)$ separate disjoint bounded intervals ought to be considered, each of which could potentially contain the solution minimizing (6.28). To minimize the reciprocal of the likelihood (6.28), we shall separately consider the three cases: The intermediate case

$$\text{Case 4.A: } X_{(r)} \leq \theta < X_{(r+1)}, \quad 1 \leq r \leq m-1, \qquad (6.30)$$

and the extreme positions:

$$\text{Case 4.B: } 0 \leq \theta < X_{(1)}, \, r = 0; \qquad (6.31)$$

$$\text{Case 4.C: } X_{(m)} \leq \theta \leq 1, r = m. \qquad (6.32)$$

Each of these cases yield a potential solution for θ for minimizing (6.28) to be denoted θ^A, θ^B and θ^C, respectively. Next, we evaluate (6.28) for these three values θ^A, θ^B and θ^C and select the one that yields the lowest value of $\mathcal{L}^{-1}(\underline{X} \mid \theta)$ given in (6.28).

The intermediate Case 4.A (Eq. (6.30): When minimizing (6.28) over $X_{(r)} \leq \theta < X_{(r+1)}$ for a specific value of r, $r = 1, \ldots, m-1$, the minimum will be attained at either $X_{(r)}$ or $X_{(r+1)}$ or at a stationary point θ^* such that $X_{(r)} \le \theta^* < X_{(r+1)}$. Introducing the notation

$$K = (n_1 - 1)r; \quad M = (n_3 - 1)(m - r), \tag{6.33}$$

denoting the function

$$g_A(\theta) = \theta^K (1 - \theta)^M \tag{6.34}$$

and setting $d\mathcal{L}^{-1}(X \mid \theta)/d\theta = 0$, we arrive at

$$((\alpha n_3 - n_1)\theta + n_1)^{m-1} \left[m g_A(\theta) + ((\alpha n_3 - n_1)\theta + n_1) g_A'(\theta) \right] = 0$$

$$\Leftrightarrow \text{either } \theta = \frac{n_1}{n_1 - \alpha n_3} \text{ or } \frac{((\alpha n_3 - n_1)\theta + n_1)}{m} = \frac{-g_A(\theta)}{g_A'(\theta)}. \tag{6.35}$$

Since for all values of $n_1, n_3, \alpha > 0$, the first solution for θ in (6.35) is less than 0 or larger than 1 it follows that only the second one can provide a proper solution $X_{(r)} \le \theta^* < X_{(r+1)}$. Hence, using the definition of $g_A(\theta)$ in (6.34) we obtain

$$\frac{((\alpha n_3 - n_1)\theta + n_1)}{m} = \frac{\theta(1 - \theta)}{K - (K + M)\theta} \Leftrightarrow$$

$$\{(\alpha n_3 - n_1)(K + M) + m\}\theta^2 - \tag{6.36}$$
$$\{m - n_1(K + M) + (\alpha n_3 - n_1)K\}\theta - n_1 K = 0.$$

Expression (6.36) is an ordinary quadratic equation possessing at most two solutions. Hence, to minimize (6.28) over $X_{(r)} \le \theta < X_{(r+1)}$, for a specific value of r, we evaluate the reciprocal of the likelihood (6.28) at $\theta_1^* = X_{(r)}$, $\theta_2^* = X_{(r+1)}$ and at the solutions θ_3^* and θ_4^* of the quadratic equation (6.36) (provided these solutions θ_3^* and θ_4^* exist and satisfy $X_{(r)} \le \theta^* < X_{(r+1)}$) and set $\theta_{(r)}$ to that value of $\theta_i^*, i = 1, \ldots 4$, which yields the minimum of $\mathcal{L}^{-1}(X \mid \theta_i^*)$ (6.28) amongst these two, three or four possibilities. Next, we evaluate the reciprocal of the likelihood $\mathcal{L}^{-1}(X \mid \theta_{(r)})$ (6.28) for $r = 1, \ldots, m - 1$ and set θ^A to be the value of $\theta_{(r)}$ that yields the minimum of $\mathcal{L}^{-1}(X \mid \theta_{(r)})$ over the set of values $r = 1, \ldots, m - 1$.

The extreme Case 4.B (see, Eq. (6.31): When minimizing (6.28) over $0 \leq \theta < X_{(1)}$, $r = 0$, the minimum is attained at either 0 or $X_{(1)}$ or at a stationary point θ^* such that $0 \leq \theta^* < X_{(1)}$. Analogously to Case 4.A additional solutions for $0 \leq \theta^* < X_{(1)}$ may be found by solving

$$\frac{((\alpha n_3 - n_1)\theta + n_1)}{m} = \frac{-g_B(\theta)}{g'_B(\theta)}$$

where

$$g_B(\theta) = (1 - \theta)^{\mathcal{M}}$$

(compare with (6.35)) and $\mathcal{M} = (n_3 - 1)m$ is given by (6.33) (recall that $r = 0$ in Case 4.B). Hence, we have

$$\theta^* = \frac{1 + n_1(n_3 - 1)}{1 + (n_1 - \alpha n_3)(n_3 - 1)}. \tag{6.37}$$

Thus, to minimize (6.28) over $0 \leq \theta < X_{(1)}$, $r = 0$, we evaluate (6.28) at $\theta_1^* = 0$, $\theta_2^* = X_{(1)}$ and $\theta_3^* = \theta^*$ given by equation (6.37) (provided that $0 \leq \theta_3^* < X_{(1)}$) and next set θ^B to be the value of θ_i^*, $i = 1, 2, 3$ which yields the minimum of $\mathcal{L}^{-1}(X \mid \theta_i^*)$, $i = 1, 2, 3$, amongst these three possibilities.

The extreme Case 4.C (see, Eq. (6.32): When minimizing (6.28) over $X_{(m)} \leq \theta \leq 1$, $r = m$, the minimum is attained at either $X_{(m)}$ or 1 or at a stationary point θ^* such that $X_{(m)} \leq \theta \leq 1$. Analogously to Case 4.A an additional solution $X_{(m)} \leq \theta^* \leq 1$ may be found by solving

$$\frac{((\alpha n_3 - n_1)\theta + n_1)}{m} = \frac{-g_C(\theta)}{g'_C(\theta)}$$

where

$$g_C(\theta) = \theta^{\mathcal{K}}$$

and $\mathcal{K} = (n_1 - 1)m$ is given by (6.33) (recall that here $r = m$). Hence, we have

$$\theta^* = \frac{n_1(n_1 - 1)}{(n_1 - \alpha n_3)(n_1 - 1) - 1}. \tag{6.38}$$

(Compare with (6.37).) Thus, to minimize the reciprocal of the likelihood (6.28) over $X_{(m)} \leq \theta \leq 1$, $r = m$, we evaluate (6.28) at $\theta_1^* = X_{(m)}$, $\theta_2^* = 1$ and $\theta_3^* = \theta^*$ given by equation (6.38) (provided that this solution θ_3^* satisfies $X_{(m)} \leq \theta_3^* \leq 1$) and set θ^C to be the value of θ_i^*, $i = 1, 2, 3$ which yields the minimum of $\mathcal{L}^{-1}(\underline{X} \mid \theta_i^*)$, $i = 1, 2, 3$, amongst these two or three possibilities.

The precondition $(n_1 \geq 1, n_3 \geq 1)$ for the ML procedure above can be relaxed to allow for the remaining parameter scenarios $(n_1 \geq 1, 0 \leq n_3 \leq 1)$, $(0 \leq n_3 \leq 1, n_3 \geq 1)$ and $(0 \leq n_1 \leq 1, 0 \leq n_3 \leq 1)$ by appropriate modifications of Step 4 (Similar to the modification of the ML estimators for unimodal STSP distribution following the proof Theorem 3.2 in Sec. 3.3 to ML estimators for U-shaped STSP distributions). The readers are encouraged to investigate these details on their own.

6.5 Illustrative Example

We shall demonstrate the ML procedure for the USTSP distribution utilizing the monthly USA Certificate Deposit rates for the period from 1966-2002. In Chapter 3 a similar example to the one discussed below used 30-year conventional mortgage interest rates over the period 1971-2003. This instructive and lengthy example is based on real-world data and involves some delicate calculations as well as several concepts of modern time series analysis which may not be familiar to some of our readers. On the other hand, it contains a few assumptions and conclusions testing their validity which may not be fully acceptable to a specialist in time series. Nevertheless we consider this carefully chosen and analyzed example to be a worthwhile contribution to a real-world statistical handling of data, providing an analysis of a situation often encountered in economic and financial applications. Our aim is to construct a realization of a time series ν_k, $k = 0, 1, 2, \ldots$, from this data, where the ν_k's are i.i.d. random variables. This would provide us with an i.i.d. sample for our ML estimation procedure. To construct such a realization we shall use the (by now quite common) Auto-Regressive Conditional Heteroscedastic (ARCH) time series model devised by R.F. Engle (a 2003 Nobel Laureate in Economics) in

1982 (which was not used in the example in Chapter 3 involving 30-year conventional mortgage interest rates).

The time series of the monthly CD rates is displayed in Fig. 6.5A consisting of 446 data points. Denoting the CD rate after month k by i_k, our starting point will be a simple financial engineering model for the random behavior of the CD rate, i.e. the multiplicative model

$$i_{(k+1)l} = i_{kl} \cdot \epsilon_{k,l}, \tag{6.39}$$

where $l = 1, 2, \ldots, 446$, $k = 0, \ldots, \lfloor 446 \cdot l^{-1} - 1 \rfloor$ and $\epsilon_{k,l}$ are i.i.d. random variables (see, e.g., Leunberger (1998)). Figure 6.5B depicts the time series of the one-step (i.e. monthly) log differences ($l = 1$ in (6.39)):

$$Ln(\epsilon_{k,1}) = Ln(i_{k+1}) - Ln(i_k) \tag{6.40}$$

totaling 445 data points and i_0 being the monthly CD rate in December of 1965. Table 6.2 contains the values of the auto-correlation function

$$ACF(\lambda, 1) = Corr[Ln(\epsilon_{k+\lambda,1}), Ln(\epsilon_{k,1})]$$

with lags $\lambda = 1, \ldots, 6$ together with the Ljung-Box Q statistics $- LBQ(\lambda) -$ (see Ljung and Box (1978)) and their p-values for testing the null hypothesis that the auto-correlations for all lags up to lag λ are zero. Tsay (2002) asserts that $\lambda \approx Ln(446) = 6.100$ performs better (as far as statistical power is concerned) than any other values. Table 6.2 contains the values of the $LBQ(\lambda)$ statistic up to $\lambda = 6$. From the corresponding p-values it follows immediately that this null hypothesis is rejected for all lags $\lambda = 1, \ldots, 6$.

Figure 6.5C depicts the time series of the two-step (i.e. bi-monthly) log differences ($l = 2$ in (6.39)):

$$Ln(\epsilon_{k,2}) = Ln(i_{2k+2}) - Ln(i_{2k}) \tag{6.41}$$

consisting of $\lfloor 446/2 - 1 \rfloor = 222$ data points where as before i_0 is the monthly CD-rate in December 1965. Table 6.2 also presents the values of the auto-correlation function

$$ACF(\lambda, 2) = Corr[Ln(\epsilon_{k+\lambda,2}), Ln(\epsilon_{k,2})]$$

with lags $\lambda = 1, \ldots, 6$ together with the corresponding $LBQ(\lambda)$ statistics and their p-values. Note that from the p-values associated with the two step

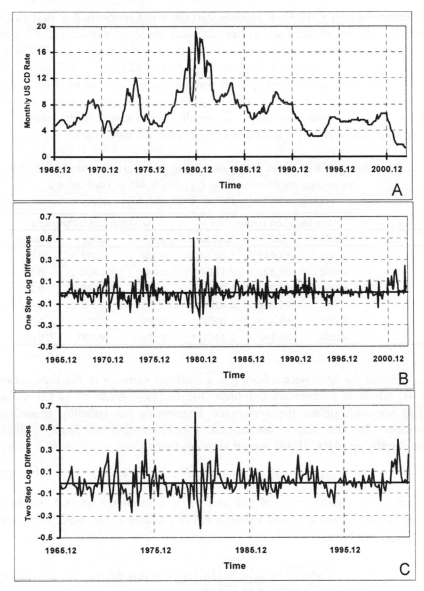

Fig. 6.5. Monthly US Certificate Deposit Rates from 1966 - 2002
A: Time Series of CD Rates B: Time Series of One-Step Log Differences;
C: Time Series of Two-Step Log Differences.

differences $Ln(\epsilon_{k,2})$ (6.41) it follows that the null hypothesis (i.e. that auto-correlations for all lags up to lag λ equal zero) is *accepted* for all lags $\lambda = 1, \ldots, 6$ (at the significance level of 4%). In the case of the 222 data points of the time series (6.41), the lag $\lambda \approx Ln(222) = 5.403$ performs the best in terms of statistical power resulting in the p-value of 0.14 in the fifth row of Table 6.2. Hence, one may reasonably conclude that the time series $Ln(\epsilon_{k,2})$ given by (6.41) is serially uncorrelated. (The reader is encouraged to analyze the situation $Ln(\epsilon_{k,3})$ for the given data.)

Table 6.2. Auto-correlation function, Ljung-Box Q statistic and p-values for one step log differences $Ln(\epsilon_{1,k})$ (Eq. (6.40)) and two-step log differences $Ln(\epsilon_{2,k})$ (Eq. (6.41)) with Lags 1,...,6.

Lag	One-Step Log Differences			Two-Step Log Differences		
	ACF	LBQ	p-value	ACF	LBQ	p-value
1	0.349	54.541	1.52E-13	0.134	4.037	0.04
2	0.092	58.321	2.17E-13	0.040	4.404	0.11
3	0.066	60.291	5.09E-13	-0.034	4.670	0.20
4	0.014	60.378	2.42E-12	0.111	7.478	0.11
5	0.007	60.401	1.00E-11	0.060	8.305	0.14
6	-0.029	60.772	3.14E-11	0.108	10.980	0.09

To test for homoscedasticity (i.e. a constant variance in the time series (6.41) which is a necessary condition for $Ln(\epsilon_{k,2})$ given by (6.41) to be i.i.d.) we shall utilize the systematic framework for volatility modeling provided by the above mentioned ARCH model of Engle (1982). Specifically, an $ARCH(m)$ model assumes that

$$a_k = \sigma_k \nu_k, \quad \sigma_k^2 = \alpha_0 + \alpha_1 a_{k-1}^2 + \ldots + \alpha_m a_{k-m}^2 \qquad (6.42)$$

where a_k is serially uncorrelated and ν_k is a sequence of i.i.d. random variables with mean zero and variance 1. For the data in Fig. 6.5C, involving $Ln(\epsilon_{k,2})$, we have

$$\overline{Ln(\epsilon_{k,2})} = \frac{1}{222} \sum_{k=0}^{221} Ln(\epsilon_{k,2}) = 0.00561; \qquad (6.43)$$

$$s^2 = \frac{1}{221}\sum_{k=0}^{221}(Ln(\epsilon_{k,2}) - \overline{Ln(\epsilon_{k,2})})^2 = 1.419e - 2.$$

(s^2 is the sample variance estimator of $Ln(\epsilon_{k,2})$, $k = 1, \ldots, 222$.) Hence, the time series

$$a_k = \frac{Ln(\epsilon_{k,2})}{s} \tag{6.44}$$

may be viewed a *realization* of (6.42) (to avoid cumbersome notation we shall use the same symbol a_k). It would seem that using (6.43) and rescaling $Ln(\epsilon_{k,2})$ as in (6.44), one achieves the conditions of a zero mean and variance 1 of ν_k in (6.42).

To further test these conditions, we present in Table 6.3 the values of the Auto-Correlation Function (ACF) and Partial Auto-correlation Function (PACF) of the time series of a_k^2 up to the lag of 5 (since $Ln(222) = 5.403$). As the first check we note that the values of the LBQ statistic (and the associated p-values) and the PACF values (in particular those in the third row) in Table 6.3 suggest that the time series a_k given by (6.44) is heteroscedastic (as opposed to homoscedastic). From the observation that a_k being serially uncorrelated (in view of (6.44) and the fact that $Ln(\epsilon_{k,2})$ are serially uncorrelated) and the PACF values of a_k^2 in Table 6.3, it follows that a_k may be well represented by an $ARCH(3)$ model (see the third row of Table 6.2 and Tsay (2002) where a detailed explanation of the theory is provided). We thus obtain the following equation for σ_t^2:

$$\sigma_k^2 = 0.6535 + 0.06657a_{k-1}^2 + 0.070194a_{k-2}^2 + 0.23382a_{k-3}^2, \tag{6.45}$$

where the parameters $\alpha = (\alpha_0, \ldots, \alpha_3)$ (Eq. (6.42)) are estimated by means of the least squares method. An alternative test for conditional heteroscedasticity is the so-called Lagrange Multiplier test also due to Engle (1982). This test is equivalent to the usual F statistic for testing $\alpha_i = 0$, $i = 1, \ldots, m$, in the linear regression (6.42) (consult a basic text on time series or Tsay (2002) for a justification of this procedure). For our data we have $F = 4.9164$ and the p-value of $2.5e - 3$ strongly confirming the earlier conclusion of heteroscedasticity of a_k as well as the $ARCH(3)$ setup as given by (6.42) and (6.45). Hence the time series a_k is <u>not</u> i.i.d..

Table 6.3. Auto-correlation function (ACF), Ljung-Box Q statistic, p-values and partial auto-correlation function (PACF) for a_k^2 (Eq. (6.44)) with Lags 1,...,5.

Lag	ACF	LBQ	p-value	PACF	t-Statistic
1	0.096	2.067	0.15	0.096	1.428
2	0.098	4.239	0.12	0.090	1.336
3	0.231	16.395	9.41E-04	0.218	3.248
4	0.098	18.581	9.50E-04	0.058	0.866
5	0.001	18.581	2.30E-03	-0.050	-0.740

However, the set-up for a_k in (6.42) and the values of σ_k given in (6.45) suggest that the standardized time series

$$\nu_k = \frac{a_k}{\sigma_k}, \ k = 3, \ldots, 221, \tag{6.46}$$

where a_k are given by (6.44), should be a realization from an i.i.d. time series (by design), which would allows us to use the standard maximum likelihood procedures. Indeed the analysis in Table 6.4 suggests that the time series ν_k is serially uncorrelated and homoscedastic — possibly with the exception of the t-statistic value of 1.423 in the fourth row of Table 6.4. This could indicate the suitability of an $ARCH(4)$ model for ν_k. However, using the linear regression to estimate the time series of σ_k^2 in (6.42) for the sequence ν_k^2, $k = 3, \ldots, 221$, results in the value of 0.8048 for the F statistic (for testing that $\alpha_i = 0$, $i = 1, \ldots, 4$) with the p-value of 0.52 which supports the homoscedasticity hypothesis of the time series (6.46). From (6.41), (6.44) and (6.46) it follows that the time series ν_k may be interpreted as that of standardized bi-monthly log-differences of US CD rates from 1966-2002; as above i_0 is the monthly CD-rate in December 1965. The empirical pdf of the standardized bi-monthly log-differences ν_k is depicted in Figs. 6.6 and 6.7 together with ML fitted Gaussian (Fig. 6.6A), asymmetric Laplace (Eq. (3.5) in Chapter 3 and Fig. 6.6B), TSP (Eq. (4.2) in Chapter 4 and Fig. 6.7A) and UTSP (Eq. (6.1) and Fig. 6.7B) distributions.

For ML estimators of the Gaussian parameters see practically any basic text in mathematical statistics, e.g., Mood *et al.* (1974). Kotz *et al.* (2002) discuss a ML estimation procedure for the asymmetric Laplace distribution in some detail. Noting the support $[-2.90, 6.97]$ of the empirical pdf ν_k depicted in Fig. 6.6, the support of the TSP and UTSP distribution in Fig. 6.7 was set (with an ample safety margin) to be $[-25, 25]$. ML estimators for the TSP distribution with a given support are provided by Eq. (4.11) in Chapter 4. Before applying the ML procedures described in Sec. 6.4 for the USTSP and GSTSP distributions, the data in Fig. 6.5C was standardized on $[0, 1]$ via a linear scale transformation applied to the original support $[-25, 25]$. Table 6.5 contains the ML estimates of the parameters of the pdf's in Figs. 6.6 and 6.7 together with those of a GTSP distribution (6.5). The parameters $\hat{\tau} \in [-25, 25]$ in Table 6.5 are obtained by applying the inverse linear scale transformation on the threshold parameter $\hat{\theta} \in [0, 1]$. The graph of the fitted GTSP distribution (6.5) is similar to that of the TSP one depicted in Fig. 6.7A.

Table 6.4. Auto-correlation function (ACF), Ljung-Box Q statistic, p-values for ν_k, and partial auto-correlation function (PACF) for ν_k^2 (Eq. (6.46)) with Lags 1,...,5.

Lag	ν_k			ν_k^2	
	ACF	LBQ	p-value	PACF	t-Statistic
1	0.122	3.322	0.07	-0.027	-0.406
2	0.099	5.490	0.06	-0.033	-0.484
3	0.060	6.307	0.10	-0.059	-0.874
4	-0.004	6.310	0.18	0.096	1.423
5	0.031	6.521	0.26	-0.017	-0.252

Analogously to the analysis in Klein (1993) (who studied interest rate data on 30-year Treasury bonds from 1977 to 1990), Fig. 6.6A shows that the empirical pdf of the financial data is by far too peaked to be captured by a normal pdf. The three-parameter asymmetric Laplace (suggested by Kozubowski and Podgórski 1999) in Fig. 6.6B is much more successful in capturing such a peak (which is a characteristic of numerous types of financial data)). From Figs. 6.6 and 6.7 we also observe (at least by a careful visual comparison) that the UTSP distribution (Fig. 6.7B) provides a "better" fit to the empirical cdf amongst these five distributions (Gaussian, Asymmetric Laplace, TSP, GTSP and UTSP). A more formal fit analysis is conducted in Table 6.6.

In Table 6.6 the Chi-square statistic

$$\sum_{i=1}^{16} \frac{(O_i - E_i)^2}{E_i} \tag{6.47}$$

189

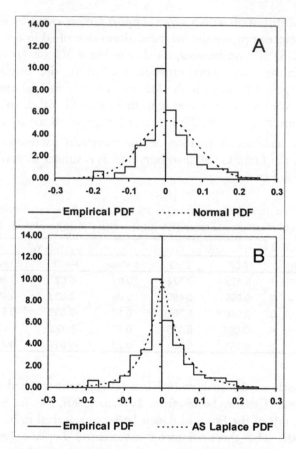

Fig. 6.6. Empirical pdf for two-step log-differences and ML fitted distributions.
A: Gaussian; B: Asymmetric Laplace.

Table 6.5. Maximum likelihood estimators for Gaussian, asymmetric Laplace, TSP,
GTSP and UTSP distributions depicted in Figs. 6.6 and 6.7.
The support for TSP, GTSP and UTSP distributions is $[-25, 25]$.

Gaussian (Fig. 6.6A)	$\hat{\mu} = 7.24e - 3$	$\hat{\sigma} = 1.095$		
AS Laplace (Fig. 6.6B)	$\hat{\mu} = -2.19e - 2$	$\hat{\sigma} = 1.033$	$\hat{\kappa} = 9.38e - 1$	
TSP (Fig. 6.7A)	$\hat{\tau} = -2.07e - 2$	$\hat{n} = 32.83$		
GTSP	$\hat{\tau} = -1.25e - 1$	$\hat{n}_1 = 38.24$	$\hat{n}_3 = 29.15$	
UTSP (Fig. 6.7B)	$\hat{\tau} = 4.65e - 2$	$\hat{n}_1 = 40.29$	$\hat{n}_3 = 25.58$	$\hat{\alpha} = 2.25$

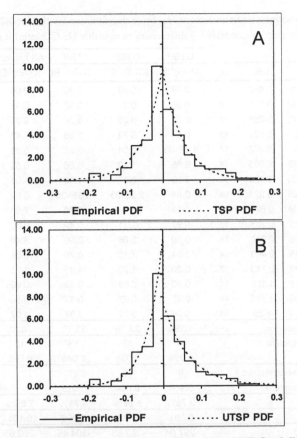

Fig. 6.7. Empirical pdf for two-step log-differences and ML fitted distributions.
A: Two-Sided Power (TSP) ; B: Uneven Two-Sided Power (UTSP).

is calculated utilizing 16 bins ($16 \in [\sqrt{219}, 219/5]$) as suggested by Banks *et al.* 2001. The boundaries of the bins are selected such that the number of observations O_i, $i = 1, \ldots, 16$, in each Bin i equals 13 or 14, totaling 219 data points. Such a boundary selection procedure partitions the support of the range of observed data in a similar manner as the "equal-probability method of constructing classes" (see, e.g., Stuart *et al.* 1994) while keeping the bin boundaries of the chi-square statistic the same across the five distributions depicted in Table 6.6. The corresponding values E_i, $i = 1, \ldots, 16$, in (6.47) for the expected number of observations in Bin i are obtained using

Table 6.6. Goodness of fit analysis of ML fitted distributions for the 1966-2002 data on standardized bi-monthly log differences of monthly US CD interest rates.

Bin	LB_i	UB_i	O_i	UTSP $(O_i\text{-}E_i)^2/E_i$	GTSP $(O_i\text{-}E_i)^2/E_i$	TSP $(O_i\text{-}E_i)^2/E_i$	AS Laplace $(O_i\text{-}E_i)^2/E_i$	Normal $(O_i\text{-}E_i)^2/E_i$
1	< -25	-1.482	13	0.77	0.36	0.30	0.06	3.27
2	-1.482	-0.913	14	0.39	0.37	0.90	0.19	3.96
3	-0.913	-0.665	14	0.01	0.05	0.04	0.27	6.3E-05
4	-0.665	-0.521	13	0.44	0.73	0.98	1.43	1.82
5	-0.521	-0.373	14	1.3E-03	0.04	0.21	0.35	1.76
6	-0.373	-0.205	14	2.05	1.30	0.56	0.52	0.49
7	-0.205	-0.125	13	0.09	0.37	1.19	1.07	9.17
8	-0.125	-0.021	14	0.84	2.8E-03	1.9E-05	0.02	5.62
9	-0.021	0.025	14	2.84	11.54	8.78	7.91	33.63
10	0.025	0.108	13	0.97	1.11	0.49	0.32	7.95
11	0.108	0.315	14	0.30	2.06	2.99	3.49	0.06
12	0.315	0.443	14	3.54	1.16	0.79	0.56	2.73
13	0.443	0.713	13	0.20	1.23	1.41	1.79	1.37
14	0.713	1.103	14	0.10	0.68	0.52	0.83	3.03
15	1.103	1.845	14	0.37	0.67	0.17	0.50	6.60
16	1.845	> 25	14	0.04	0.52	3.37	1.07	0.11
Chi-Squared Statistic				12.96	22.19	22.70	20.38	81.57
Degrees of Freedom				11	12	13	12	13
P-value				0.296	0.035	0.045	0.060	< 1.00E-6
Degrees of Freedom (Disc.)				9	10	11		
P-value (Discounted)				0.164	0.014	0.019		
K-S Statistic				3.76%	5.58%	4.71%	7.65%	12.75%
SS				0.051	0.105	0.068	0.194	1.126
Log-Likelihood				-297.04	-302.83	-304.45	-302.55	-332.39

$$E_i = 219\{F(UB_i|\widehat{\Theta}) - F(LB_i|\widehat{\Theta})\},$$

where $F(\cdot|\widehat{\Theta})$ is the theoretical cdf, $\widehat{\Theta}$ are the ML estimators for the parameters given in Table 6.5 for each distribution and the bin boundaries (LB_i, UB_i) are presented in Table 6.6. Evidently the Gaussian distribution produces the worst fit with 12 out of the 16 bins contributing a value 1.00 or more to the chi-squared statistic (6.47). In particular the very high value 33.63 for Bin 9, containing the peak in the empirical pdf, reconfirms the conclusion obtained from Fig. 6.6A that the Gaussian distribution in no way represents such a "peak". While the other fitted distributions (UTSP,

GTSP, TSP and asymmetric Laplace) perform much better from bin to bin, Bin 9 by far contributes the most to the chi-squared statistics regardless of the type of distribution, except for the UTSP case (where it provides the second largest value).

The UTSP distribution yields a better value in terms of the chi-squared statistic not only due to a substantial smaller value in Bin 9, but also because the remaining bins in the UTSP case contribute in total the least to the overall value of the chi-squared statistic compared with the other four distributions, thus overall more correctly tracing the empirical distribution. In addition, the UTSP distribution results in the largest p-value of the chi-squared hypothesis test taking into account the number of parameters of each distribution to determine the degrees of freedom. This observation also applies to the second pair of p-values and degrees of freedom calculated by discounting two additional degrees of freedom for the boundary parameters $[-25, 25]$. (It is not quite clear that these two degrees of freedom should be discounted since the boundaries -25 and 25 were not formally estimated from the data but rather obtained by observation.)

Table 6.6 also includes the popular Kolmogorov-Smirnov Statistic D (see, e.g., Stuart *et al.* 1994)

$$D = Max\{D_i | i = 1, \ldots, 219\} \tag{6.48}$$

where

$$D_i = Max\left\{|\frac{i-1}{219} - F(X_{(i)}|\widehat{\Theta})|, |\frac{i}{219} - F(X_{(i)}|\widehat{\Theta})|\right\}, \tag{6.49}$$

as well as "an intuitive measure of fit"

$$\sum_{i=1}^{219}\left(\frac{i}{219} - F(X_{(i)}|\widehat{\Theta})\right)^2, \tag{6.50}$$

denoted by Sum of Squares (SS) (reminiscent of the sum of squares in linear regression analysis) and the log-likelihood

$$\sum_{i=1}^{219} Ln\{f(X_{(i)}|\widehat{\Theta})\} \tag{6.51}$$

where in (6.49), (6.50) and (6.51), $X_{(i)}$, $i = 1, \ldots, 219$, are the order statistics associated with the standardized bi-monthly log-differences ν_k (see, (6.46)) and as mentioned above the vector $\widehat{\Theta}$ consists of the ML estimators of the parameters given in Table 6.5. Note that the UTSP distribution performs best for all the statistics (6.47), (6.48), (6.50) and (6.51) amongst the five distributions presented in Table 6.6. Somewhat unexpectedly, the GTSP and TSP distributions outperform the asymmetric Laplace distribution in terms of the Kolmogorov-Smirnov statistic (6.48) and the SS (6.50), but not in terms of the chi-squared statistic (6.47) (and its p-value) and the log-likelihood (6.51) of the data involving ν_k. The K-S statistic and the log-likelihood seem to be much less sensitive to the evident inappropriateness of the Gaussian distribution in the situation at hand. The results obtained require of course further elaboration by analyzing the structure of data representation and taking into account assumptions under which the selected test statistics are appropriate. In the authors' opinion the behavior of the chi-squared statistic in the pivotal Bin 9 in Table 6.6 (corresponding to the values in the vicinity of the "peak") strongly justifies the conclusion about the suitability of the UTSP distribution for the data under consideration.

Chapter 7

The Reflected Generalized
Topp and Leone Distribution

Before reading this chapter the reader is advised to review the material of the first two chapters. Here we shall present a new two parameter family of continuous distribution on a bounded domain which has an elevated but finite density value at its lower bound. Such a situation is encountered, for example, when representing income distributions at lower income ranges. The new family generalizes the one parameter Topp and Leone distribution originated in the 1950's which was discussed in Chapter 2, Sec. 2.1. It is connected to the rather involved generalized trapezoidal distributions discussed in Chapter 5 via the slope distribution (5.6) and the reflected power distribution (5.5) (see also Fig. A.1 in Appendix A). Both the slope and the reflected power distributions served as important building blocks for the generalization of the trapezoidal distribution. While the family of beta distributions has been used extensively for modeling bounded income distribution situations, it allows only for an infinite or zero density values at its lower bound with a strictly positive mode or anti-mode at the same time (and a constant density of 1 in case of its uniform member). The proposed new family alleviates this apparent jump discontinuity at the lower bound. The U.S. income distribution data for the year 2001 is used to fit distributions for Caucasian (Non-Hispanic), Hispanic and African-American populations via a maximum likelihood procedure. The results reveal a proper stochastic ordering when comparing the Caucasian (Non-Hispanic) income distribution to that of the Hispanic or African-American population. The existence of this ordering indicates that although substantial advances have reportedly been made in reducing the income distribution gap amongst the different ethnic groups in the U.S. population during the last 20 years since the eighties of the past century, these differences still existed in the early years of the 21-st century.

7.1 Introduction

As it was alluded to in Chapter 3, in an early issue of the *Journal of the American Statistical Association*, in 1955, before the appearance of computer assisted statistical methodology, an isolated paper on a bounded continuous distribution with the pdf (2.6) has appeared co-authored by C. Topp and F. Leone which originally received little attention. While the Topp and Leone distribution utilized the left triangular distribution with the pdf $2 - 2x$, $x \in (0, 1)$ (depicted in Fig. 2.2A in Chapter 2) as its *generating density*, Van Dorp and Kotz's (2004b) generalization of the Topp and Leone distribution (GTL) utilizes a slightly more general generating pdf

$$\alpha - 2(\alpha - 1)x, \ 0 \leq \alpha \leq 2, \ x \in (0, 1) \tag{7.1}$$

(see Fig. 7.1A with $\alpha = 1.5$). From the restriction that $\alpha - 2(\alpha - 1)x \geq 0$ for all $x \in (0, 1)$, it follows that $0 \leq \alpha \leq 2$. For the extreme value $\alpha = 2$, the generating pdf (7.1) reduces to the left triangular pdf in Fig. 2.2A. We shall refer to distributions with the pdf of the form (7.1) as *slope distributions*. A slope distribution possesses a linear pdf and also plays a central role in deriving a generalization of the trapezoidal distribution (see Eq. (5.6) in Chapter 5, which is in a reparameterized form of the pdf (7.1)). For $\alpha \in (1, 2]$ ($\alpha \in [0, 1)$), the slope of the pdf (7.1) is decreasing (increasing). For $\alpha = 1$, the slope distribution reduces to a uniform distribution on $(0, 1)$.

Figure 7.1B plots the *generating* cdf $\alpha x - (\alpha - 1)x^2$ for $\alpha = 1.5$ associated with the linear pdf presented in Fig. 7.1A. Now the *generalized* Topp and Leone (GTL) distribution that follows from Fig. 7.1B utilizing the construction described by Eq. (2.1) in Chapter 2 is depicted in Fig. 7.1D for the value $\beta = 3$. The density associated with the cdf given by

$$\{\alpha x - (\alpha - 1)x^2\}^3$$

(compare with Eq. (2.4)) with $\alpha = 1.5$ is displayed in Fig. 7.1C. Note that, while a mode in $(0, 1)$ is still present in Fig. 7.1C, it has been shifted to the right. More importantly, the density at the upper bound is strictly positive in Fig. 7.1C instead of being zero in Fig. 7.2C (which represents the original Topp and Leone density).

Our illustrative example in this chapter (Sec. 7.5) involves income data for the US households in the year of 2001. For this purpose we shall

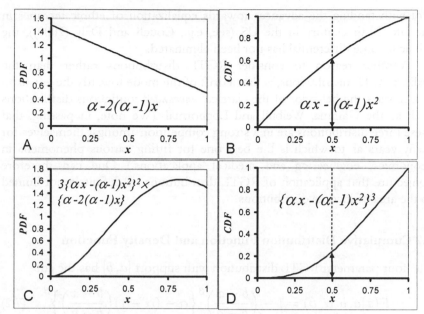

Fig. 7.1 Construction of generalized Topp and Leone distribution from a slope distribution.
A: Slope pdf with $\alpha = 1.5$; B: The corresponding cdf of the pdf in Fig. 7.1A;
C: Generalized Topp and Leone pdf with $\alpha = 1.5$, $\beta = 3$;
D: The corresponding cdf of the pdf in Fig. 7.1C.

consider the *reflected version* of the Generalized Topp and Leone (GTL) distribution utilizing the cdf transformation $H(x) = 1 - G(1 - x)$, where G is a GTL cdf on $[0, 1]$. The transformation typically assigns the mode towards the left-hand-side of its support and allows for strictly positive density values at the lower bound. This form seems to be appropriate when representing income distributions at lower income ranges. (Compare, e.g., with Fig. 2 of Barsky *et al.* (2002), p. 668.) The U.S. Income distribution data for the year 2001 is used to fit *reflected* GTL (RGTL) distributions for the household incomes of Caucasian (Non-Hispanic), Hispanic and African-American populations using a maximum likelihood procedure. The results reveal stochastic ordering when comparing the Caucasian (Non-Hispanic) income distribution to that of the Hispanic or African-American populations. Compared to Americans of Caucasian origin, African-Americans appear to be approximately 1.9 times as likely and the Hispanics 1.5 times as likely to have inadequate or no income at all.

197

Not withstanding the advances towards equalization of ethnic incomes in the late 20-th century in the US (see, e.g., Couch and Daly (2000)), the ethnic income differential has not been eliminated.

Another reason to consider RGTL distributions rather than the ordinary GTL distributions, is that a drift of the mode towards the left-hand side mimics the behavior of the classical *unbounded* continuous distributions such as the Gamma, Weibull and Lognormal. (We note, in passing, that these three distributions are in a strong competition amongst themselves for many years as to which is the best one for fitting various phenomena in economics, engineering and medical applications.) One can therefore conjecture that application of RGTL distributions is definitely not limited to the area of income distributions.

7.2 Cumulative Distribution Function and Density Function

The four parameter RGTL distribution with support $[a, b]$ has the cdf

$$F(x|a,b,\alpha,\beta) = 1 - \left(\frac{b-x}{b-a}\right)^{\beta}\left\{\alpha - (\alpha-1)\left(\frac{b-x}{b-a}\right)\right\}^{\beta} \quad (7.2)$$

where

$$a \leq x \leq b, \ 0 < \alpha \leq 2 \text{ and } \beta > 0. \quad (7.3)$$

Recall that α is the parameter of the generating slope distribution. Evidently, the cdf value at the lower bound a (upper bound b) equals zero (one). The pdf follows from (7.2) to be

$$f(x|a,b,\alpha,\beta) = \frac{\beta}{b-a}\left(\frac{b-x}{b-a}\right)^{\beta-1} \times \quad (7.4)$$

$$\left\{\alpha - (\alpha-1)\left(\frac{b-x}{b-a}\right)\right\}^{\beta-1}\left\{\alpha - 2(\alpha-1)\left(\frac{b-x}{b-a}\right)\right\},$$

with the same constraints (7.3) on x, α and β (compare with Eq. (2.6) in Chapter 2). From Eq. (7.4) it follows that, in particular, at the endpoints the density becomes

$$f(a|a,b,\alpha,\beta) = \frac{\beta(2-\alpha)}{b-a} \quad (7.5)$$

and

$$f(b|a, b, \alpha, \beta) = \begin{cases} 0 & \beta > 1 \\ \frac{\beta\alpha}{b-a} & \beta = 1 \\ \to \infty \text{ as } x \uparrow b & \beta < 1. \end{cases} \tag{7.6}$$

Equation (7.5) shows that the RGTL family allows for arbitrary density values at the lower endpoint a. The standardized versions of the TL pdf and cdf follow by substituting $b = 1$ in Eqs. (2.6) and (2.7) and are given by Eqs. (2.5) and (2.4) in Chapter 2, respectively. Figure 2.3C in Chapter 2 depicts a graph of a standard TL distribution with parameters $b = 1$ and $\beta = 3$ in Eq. (2.6). Figure 7.2A displays its reflected version.

Note the transition of the form of graphical representations of the pdf's from Fig. 7.2B to Fig. 7.2D via Fig. 7.2C which all have the same value of α but decreasing values of β. Observe that in case of Fig. 7.2B the pdf assumes a similar form to that of the reliability function $1 - F(\cdot)$ whereas Fig. 7.2C displays a mode at a value greater than 0. Similarly in Figs. 7.2E to 7.2H the pdf's with the same value of α ($= 0.5$) with progressively decreasing β from 2 to 0.25 change its form from a monotonically decreasing concave, via a linear function with decreasing slope, followed by a mild U-shaped function, up to a monotonically increasing convex curve. The J-shaped form of the pdf in Fig. 7.2E ($a = 0, b = 1, \alpha = 0.5, \beta = 2$) resembles that of a Weibull distribution with the shape parameter less than one (but on a bounded domain). Note that the structure of the cdf (7.2) is reminiscent to that of a Weibull cdf. Figs. 7.2G and 7.2H depict a U-shaped form ($a = 0, b = 1, \alpha = 0.5, \beta = 0.75$) and a J-shaped form ($a = 0, b = 1, \alpha = 0.5, \beta = 0.25$), respectively, and are similar to those appearing in the beta family of distributions, but consistently with a *bounded density value* at its lower bound (Eq. (7.5)).

Evidently, setting $\alpha = 1, \beta = 1$ into the pdf (7.4) yields a uniform distribution on $[a, b]$. Hence, analogously to the four parameter beta distribution with the pdf (4.1) in Chapter 4 and the TSP distribution with the pdf (4.2), the RGTL family has the uniform distribution on $[a, b]$ as one of its members. Another common member amongst these three families (Beta, TSP and RGTL) is the <u>reflected</u> power (RP) distribution on $[a, b]$ with the pdf

$$f(x|a, b, 1, \beta) = \frac{\beta}{b - a}\left(\frac{b - x}{b - a}\right)^{\beta-1} \tag{7.7}$$

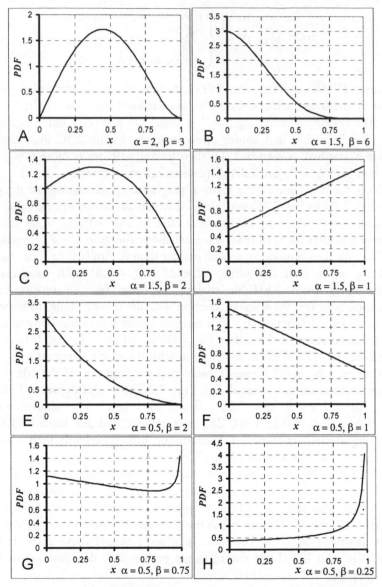

Fig. 7.2 Examples of standard ($a = 0$, $b = 1$) RGTL distributions:
A: $\alpha = 2$, $\beta = 3$; B: $\alpha = 1.5, \beta = 6$; C: $\alpha = 1.5, \beta = 2$; D: $\alpha = 1.5, \beta = 1$;
E: $\alpha = 0.5, \beta = 2$; F: $\alpha = 0.5, \beta = 1$; G: $\alpha = 0.5, \beta = 0.75$; H: $\alpha = 0.5, \beta = 0.25$.

obtained by substituting $\alpha = 1$ in the pdf (7.4) (namely utilizing the uniform distribution as its generating density). Substituting $\alpha = 0$ in (7.4) also yields the reflected power distribution (applying the left triangular distribution with pdf $2x$ on $[0, 1]$ as its generating density) but with parameter 2β. The reader is encouraged to study Fig. A.1 in Appendix A which connects the above cited distributions.

In addition to the distinguishing feature of the RGTL family − the existence of pdf's with positive density value at the lower limit with a strict positive mode − another feature of RGTL distribution (indicating possibly a lesser flexibility within the same family) is that the pdf's of a GTL distributions and its reflections possess *different* functional forms, while the reflections of TSP and beta pdf's belong to the same functional family.

7.3 Properties of Standard RGTL Distributions

We shall list some properties of the Standard RGTL (SRGTL) distributions (setting $a = 0$ and $b = 1$ in (7.2) and (7.4)) with the cdf

$$F(x|\alpha, \beta) = 1 - (1 - x)^{\beta}\left\{\alpha - (\alpha - 1)(1 - x)\right\}^{\beta} \qquad (7.8)$$

and the pdf

$$f(x|\alpha, \beta) = \beta(1 - x)^{\beta-1} \times \qquad (7.9)$$
$$\left\{\alpha - (\alpha - 1)(1 - x)\right\}^{\beta-1}\left\{\alpha - 2(\alpha - 1)(1 - x)\right\},$$

where $0 < \alpha \leq 2$ and $\beta > 0$. Results may be directly extended to the general forms of (7.2) and (7.4) by means of a simple linear transformation.

7.3.1 Limiting distributions

It immediately follows from (7.8) that as $\beta \to \infty$ ($\beta \downarrow 0$) the pdf (7.9) converges to a degenerate distribution with a probability mass of 1 at 0 (at 1) since $F(x|\alpha, \beta) \to 1$ for $x \in (0, 1]$ (since $F(x|\alpha, \beta) \to 0$ for $x \in [0, 1)$) regardless of the value of α.

7.3.2 Some stochastic dominance properties

Note that for $\beta = 1$ the cdf of a SRGTL distribution (7.8) reduces to a slope distribution with the cdf

$$F(x|\alpha, \beta = 1) = 1 - \{\alpha(1 - x) - (\alpha - 1)(1 - x)^2\} \qquad (7.10)$$

which is stochastically decreasing in α, i.e.,

$$\alpha_1 \leq \alpha_2, x \in (0, 1) \Rightarrow F(x|\alpha_1, \beta = 1) \geq F(x|\alpha_2, \beta = 1). \qquad (7.11)$$

Let now $\beta_1 \geq \beta_2 > 0$. From (7.11) it follows that for all $x \in (0, 1)$ and for any β_1

$$1 - \{1 - F(x|\alpha_1, \beta = 1)\}^{\beta_1} \geq 1 - \{1 - F(x|\alpha_2, \beta = 1)\}^{\beta_1}. \qquad (7.12)$$

Since the function z^a is a decreasing function in a for $z \in (0,1)$ it follows from $\beta_1 \geq \beta_2 > 0$ that

$$1 - \{1 - F(x|\alpha_2, \beta = 1)\}^{\beta_1} \geq 1 - \{1 - F(x|\alpha_2, \beta = 1)\}^{\beta_2}. \qquad (7.13)$$

However, simple algebra shows that

$$F(x|\alpha, \beta) = 1 - \{1 - F(x|\alpha, \beta = 1)\}^{\beta}$$

where $F(x|\alpha, \beta)$, $F(x|\alpha, \beta = 1)$ are given by (7.8) and (7.10), respectively, which together with (7.12) and (7.13) implies

$$\alpha_1 \leq \alpha_2, \beta_1 \geq \beta_2, x \in (0, 1) \Rightarrow F(x|\alpha_1, \beta_1) \geq F(x|\alpha_2, \beta_2). \qquad (7.14)$$

Hence, RGTL distributions are stochastically increasing in α and stochastically decreasing in β. This seems to be an interesting property shedding an additional light on the meaning of the parameters α and β in (7.8) and (7.9), especially in applications. Note that relation (7.14) could verbally be expressed as connecting the *generating* cdf $F(x|\alpha, \beta = 1)$ with the *generated* one, i.e. $F(x|\alpha, \beta)$. The reader is invited to check other stochastic dominance properties of other continuous distributions (not necessarily on a bounded domain).

7.3.3 Mode analysis of standard RGTL distributions

As it was already mentioned for $\beta = 1$ and $\alpha = 1$ the pdf (7.9) reduces to a uniform $[0, 1]$ density. For $\alpha = 1$, $\beta \neq 1$ the pdf (7.9) reduces to a reflected power distribution with a finite mode at 0 provided $\beta > 1$ and an infinite mode at 1 for $\beta < 1$. Taking the derivative of (7.9) with respect to x we have

$$\frac{df(x|\alpha, \beta)}{dx} = C(x|\alpha, \beta)f(x|\alpha, \beta), \tag{7.15}$$

where the multiplier

$$C(x|\alpha, \beta) = (\alpha - 1)\frac{2}{\alpha - 2(\alpha - 1)\left(1 - x\right)} - \tag{7.16}$$

$$(\beta - 1)\frac{\left\{\alpha - 2(\alpha - 1)(1 - x)\right\}}{(1 - x)\left\{\alpha - (\alpha - 1)(1 - x)\right\}}$$

is a linear function in β. From the relations

$$f(x|a, b, \alpha, \beta) > 0, \tag{7.17}$$
$$\{\alpha - 2(\alpha - 1)(1 - x)\} > 0 \text{ and}$$
$$\{\alpha - (\alpha - 1)(1 - x)\} > 0$$

for $\alpha \in [0, 2]$, and $\beta > 1$ it follows from (7.15) and (7.16) that the following four additional cases should be considered:

Case 1 : $0 < \alpha < 1, \beta \geq 1$;
Case 2 : $1 < \alpha \leq 2, \beta \leq 1$;
Case 3 : $1 < \alpha \leq 2, \ \beta > 1$ and
Case 4 : $0 < \alpha < 1, \ \beta < 1$.

Case 1: $0 < \alpha < 1, \beta \geq 1$: See Figs. 7.2E and 7.2F : From (7.15), (7.16) and (7.17) it follows that the SRGTL pdf (7.9) is strictly decreasing on $[0, 1]$ and hence possesses a mode at 0 with the value $\beta(2 - \alpha)$ (Eq. 7.5)). For example, setting $\alpha = 0.5$ and $\beta = 2$ (as in Fig. 7.2E) yields a mode at 0 with value 3. Setting $\alpha = 0.5$ and $\beta = 1$ (as in Fig. 7.2F) yields a mode at 0 with value 1.5.

$\underline{\text{Case 2:}}$ $1 < \alpha \leq 2, \beta \leq 1$: See Fig. 7.2D : From (7.15), (7.16) and (7.17) it follows that the SRGTL pdf (7.9) is in this case strictly *increasing* on $[0, 1]$. From (7.6) it follows that the pdf (7.9) has an infinite mode at 1 for $\beta < 1$ and a finite mode at 1 for $\beta = 1$. Setting $\alpha = 1.5$ and $\beta = 1$ (as in Fig.7.4D) yields a finite mode at 1 with value 1.5.

$\underline{\text{Case 3:}}$ $1 < \alpha \leq 2, \beta > 1$: See Figs. 7.2A, 7.2B and 7.2C : This seems to be the most interesting case with the pdf vanishing at $x = 1$. From (7.15), (7.16) and (7.17) it follows that the SRGTL pdf (7.9) may possess a mode in $(0, 1)$. Defining

$$y = 1 - x$$

and setting the derivative (7.15) to zero yields the following quadratic equation in y:

$$2(\alpha - 1)^2 y^2 - 2\alpha(\alpha - 1)y + \frac{\alpha^2(\beta - 1)}{2\beta - 1} = 0. \qquad (7.18)$$

(The left-hand-side (LHS) of (7.18) is a parabolic function in y.) Noting that the symmetry axis of the parabola associated with the LHS of (7.18) has the value

$$\frac{\alpha}{2(\alpha - 1)} \qquad (7.19)$$

which is strictly greater than 1 for $\alpha > 1$, and that $y = 1 - x \in [0, 1]$ if and only if $x \in [0, 1]$, it follows that out of the two possible solutions of the quadratic equation (7.18) only the solution

$$y^* = \frac{\alpha}{2(\alpha - 1)} \cdot \left\{ 1 - \sqrt{\frac{1}{2\beta - 1}} \right\} \qquad (7.20)$$

can yield a mode $x^* \in (0, 1)$. Moreover, from $1 < \alpha \leq 2$ and $\beta > 1$ it follows that $y^* > 0$. Also, from (7.20) we have that $y^* \to \frac{\alpha}{2(\alpha-1)} > 1$ for $1 < \alpha \leq 2$ as $\beta \to \infty$. Hence, from (7.20) we deduce that the mode $x^* = 1 - y^*$ is

$$x^* = Max\left[0, \frac{1}{2(\alpha - 1)}\left\{\alpha\left(1 + \sqrt{\frac{1}{2\beta - 1}}\right) - 2\right\}\right]. \quad (7.21)$$

Setting $\alpha = 1.5, \beta = 2$ (as in Fig. 7.2C) yields $x^* = Max[0, -\frac{1}{2} + \frac{1}{2}\sqrt{3}]$ ≈ 0.366. As β increases the mode moves to the left eventually approaching zero. Indeed, setting $\alpha = 1.5$, $\beta = 6$ (as in Fig. 7.2B) yields $x^* = Max[0, -\frac{1}{2} + \frac{3}{22}\sqrt{11}] = 0$ and hence a mode is located at the lower bound 0 with the value $\beta(\alpha - 2) = 3$ (Eq. (7.5) with $a = 0$, $b = 1$). Utilizing (7.21) it follows that a Standard Reflected Topp and Leone (SRTL) distribution (with $\alpha = 2$) has a mode at

$$\sqrt{\frac{1}{2\beta - 1}} \quad (7.22)$$

for $\beta \geq 1$. Setting $\beta = 3$ (as in Fig. 7.1A) yields a mode at $\frac{1}{5}\sqrt{5} \approx 0.447$.

Case 4: $0 < \alpha < 1$, $\beta < 1$: See Figs. 7.2G and 7.2H : Similarly to Case 2 it follows that the pdf (7.9) has an infinite mode at 1 for $0 < \alpha < 1$, $\beta < 1$. However, from (7.15), (7.16) and (7.17) it follows that the pdf (7.9) may also have an anti-mode $x^* \in (0, 1)$ (resulting in a U-shaped distribution). In fact, the formula for the anti-mode is also given by (7.21) provided $\beta > \frac{1}{2}$. For example, setting $\alpha = 0.5$, $\beta = 0.75$ (as in Fig. 7.2G) yields $x^* = Max[0, \frac{3}{2} - \frac{1}{2}\sqrt{2}]$ and hence the anti-mode is at approximately 0.793. For $\beta \leq \frac{1}{2}$ (as in Fig. 7.2H) the anti-mode of an RGTL distribution occurs at $x^* = 0$, with density value (Eq. (7.5) with $a = 0$, $b = 1$)

$$\beta(2 - \alpha) \leq 1 - \frac{\alpha}{2} \leq 1. \quad (7.23)$$

In Fig. 7.2H the density value at its antimode equals $3/8$.

The authors consider Sec. 7.3.3 to be one of their favorite ones. The remarkable journey walk of the mode as a function of its parameters α and β shows the flexibility and power of analytical closed representation of density functions (even most elementary ones).

7.3.4 Failure rate function

From (7.8) and (7.9) it follows that the failure rate function

$$r(t) = \frac{f(t)}{1 - F(t)}$$

(also known as the hazard rate function) for an SRGTL density is :

$$\mathcal{D}(\alpha, x)\frac{\beta}{1 - x} \qquad (7.24)$$

where

$$\mathcal{D}(\alpha, x) = \frac{\alpha - 2(\alpha - 1)(1 - x)}{\alpha - (\alpha - 1)(1 - x)} \qquad (7.25)$$

and it is straightforward to verify that $\beta/(1 - x)$ is the failure rate of a standard reflected power (SRP) distribution (Eq. (7.9) with $\alpha = 1$). From (7.24) one observes that $\mathcal{D}(\alpha, x)$ can be interpreted as the relative increase (or decrease) in the failure rate of a SRGTL distribution as compared to a SRP distribution. Taking the derivative of (7.25) with respect to x we obtain

$$\frac{\partial \mathcal{D}(\alpha, x)}{\partial x} = \frac{\alpha(1 - \alpha)}{\{\alpha - (\alpha - 1)(1 - x)\}^2}. \qquad (7.26)$$

Hence, $D(1, x) = 1$ for all $x \in [0, 1\}$ and it follows from (7.26) that $D(\alpha, x) < 1$ (> 1) for all $x \in [0, 1]$ as long as $1 < \alpha \le 2$ ($0 \le \alpha < 1$). Thus, α may be interpreted as a *failure deceleration parameter* (relative to the SRP distribution) when $1 < \alpha \le 2$ and a failure acceleration parameter when $0 \le \alpha < 1$. Namely, a larger value of α causes a smaller failure rate than that of the reflected power distribution and vice versa. On the other hand, (7.24) shows that β is a *failure acceleration parameter* for all $\beta > 0$. Namely, a larger value of β results (continuously) in a larger failure rate.

7.3.5 Cumulative moments

Due to the functional form of the cdf (7.8) calculations of the cumulative moments defined by

$$M_k = \int_0^1 x^k (1 - F(x)) dx \tag{7.27}$$

for SRGTL distributions seems to be more direct than evaluation of the central moments about the mean. The mean μ_1' and the central moments about the mean: i.e. μ_2 (variance), μ_3 (non-standardized skewness) and μ_4 (non-standardized kurtosis) are connected with the cumulative moments $M_k, k = 1, \ldots, 4$, via the relations

$$\mu_1' = M_0 \tag{7.28}$$
$$\mu_2 = 2M_1 - M_0^2$$
$$\mu_3 = 3M_2 - 6M_1 M_0 + 2M_0^3$$
$$\mu_4 = 4M_3 - 12M_2 M_0 + 12M_1 M_0^2 - 3M_0^4$$

(see, e.g., Stuart and Ord (1994)). Consequently, the coefficient of skewness β_1 and the coefficient of kurtosis β_2 defined as

$$\beta_1 = \frac{\mu_3^2}{\mu_2^3}, \ \beta_2 = \frac{\mu_4}{\mu_2^2} \tag{7.29}$$

can be straightforwardly obtained via the results (7.28).

The k-th cumulative moment M_k for SRGTL distributions is (utilizing Eqs. (7.8) and (7.27)):

$$\int_0^1 x^k (1-x)^\beta \Big\{ \alpha - (\alpha - 1)(1 - x) \Big\}^\beta dx = \tag{7.30}$$
$$\sum_{i=0}^k \binom{k}{i} (-1)^i \alpha^\beta \int_0^1 x^{\beta+i} \Big\{ 1 - \frac{(\alpha-1)x}{\alpha} \Big\}^\beta dx$$

For $\alpha = 1$, Eq. (7.30) reduces to that of the k-th cumulative moment of a SRP distribution (Eq. (7.9) with $\alpha = 1$). For $\alpha \in (1, 2]$, the k-th cumulative moment can be expressed in terms of the incomplete Beta function

$$B(x \mid a, b) = \frac{\Gamma(a + b)}{\Gamma(a)\Gamma(b)} \int_0^x p^{a-1}(1 - p)^{b-1} dp \tag{7.31}$$

as

$$M_k = \sum_{i=0}^{k} \binom{k}{i}(-1)^i \alpha^\beta \left\{ \frac{\alpha}{\alpha - 1} \right\}^{\beta+i+1} \times \tag{7.32}$$

$$\left\{ \frac{\Gamma(\beta + i + 1)\Gamma(\beta + 1)}{\Gamma(2\beta + i + 2))} B(\frac{\alpha - 1}{\alpha} \mid \beta + i + 1, \beta + 1) \right\}.$$

Numerical routines for evaluating the incomplete Beta function (7.31) are by now well known and are provided in standard PC software such as Microsoft EXCEL. However, for $\alpha \in (0, 1)$ expression (7.30) cannot be further simplified and one has to resort to numerical integration.

For $\alpha \in (1, 2]$, we have for the cumulative moments M_0, M_1, M_2 and M_3 of SRGTL distribution to be:

$$\mu_1' = M_0 = \alpha^\beta \left\{ \frac{\alpha}{\alpha - 1} \right\}^{\beta+1} \left\{ \frac{B(\frac{\alpha-1}{\alpha} \mid \beta + 1, \beta + 1)}{\mathbb{B}^{-1}(\beta + 1, \beta + 1)} \right\} \tag{7.33}$$

$$M_1 = M_0 - \alpha^\beta \left\{ \frac{\alpha}{\alpha - 1} \right\}^{\beta+2} \left\{ \frac{B(\frac{\alpha-1}{\alpha} \mid \beta + 2, \beta + 1)}{\mathbb{B}^{-1}(\beta + 2, \beta + 1)} \right\}$$

$$M_2 = -M_0 + 2M_1 + \alpha^\beta \left\{ \frac{\alpha}{\alpha - 1} \right\}^{\beta+3} \left\{ \frac{B(\frac{\alpha-1}{\alpha} \mid \beta + 3, \beta + 1)}{\mathbb{B}^{-1}(\beta + 3, \beta + 1)} \right\}$$

$$M_3 = M_0 - 3M_1 + 3M_2 - \alpha^\beta \left\{ \frac{\alpha}{\alpha - 1} \right\}^{\beta+4} \left\{ \frac{B(\frac{\alpha-1}{\alpha} \mid \beta + 4, \beta + 1)}{\mathbb{B}^{-1}(\beta + 4, \beta + 1)} \right\},$$

where the beta function $\mathbb{B}(a,b) = \Gamma(a)\Gamma(b)/\Gamma(a + b)$. For the original Topp and Leone (1955) distribution (Eq. (2.6) in Chapter 2) the cumulative moments were derived by Nadarajah and Kotz (2003). Substituting $\alpha = 2$, in (7.33) yields the mean

$$\mu_1' = M_0 = 4^\beta \mathbb{B}(\beta + 1, \beta + 1)$$

of a Standard Reflected Topp and Leone (SRTL) distribution and hence

$$\mu_1' = M_0 = 1 - 4^\beta \mathbb{B}(\beta + 1, \beta + 1)$$

is the mean of a Standard Topp and Leone (STL) distribution on $(0, 1)$ (see Nadarajah and Kotz (2003) and Eq. (2.15) in Chapter 2).

Figure 7.3 plots the behavior of the mean, variance, skewness and kurtosis (as a function of the parameter β) for a SRP ($\alpha = 1$ in (7.9)) distribution, a SRGTL distribution with $\alpha = 1.5$ in (7.9) and a Standard Reflected Topp and Leone (SRTL) distribution ($\alpha = 2$ in (7.9)).

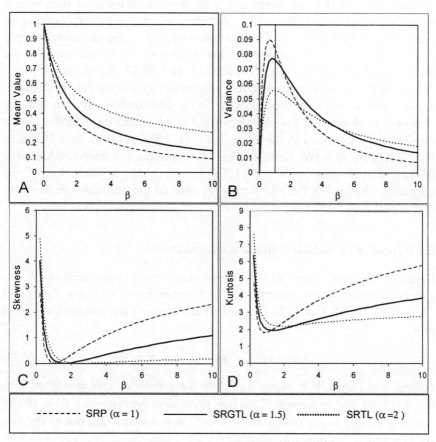

Fig. 7.3 Behavior of the mean (A), variance (B), skewness (C) and kurtosis (D) of a standard reflected power ($\alpha = 1$ in (7.9)), a standard generalized reflected Topp and Leone ($\alpha = 1.5$ in (7.9)) and a standard reflected Topp and Leone ($\alpha = 2$ in (7.9)) distributions.

Below we shall briefly comment on the behavior of the characteristics in Fig. 7.3. Figure 7.3A shows that for $\beta \downarrow 0$ the *mean values* all have limiting value of 1 due to a degenerate limiting distribution with a probability mass of 1 at 1 (see Sec. 7.3.1). Also note the strict ordering in the mean values in Fig. 7.3A for fixed values of β. This ordering also follows immediately from the stochastic dominance property (7.14) resulting in larger values for SRTL ($\alpha = 2$) and the smallest for SRP ($\alpha = 1$). The variance (Fig. 7.3B) has a limiting value of 0 when either $\beta \downarrow 0$ or $\beta \to \infty$. Also note that, with a possible exception of the SRTL distribution, the maximal variance is attained for values of $\beta < 1$. For the values of $\beta > 1$, the variance is strictly decreasing. Figure 7.3C indicates that skewness of 0 are attained for different values of β for the SRP, SRGTL and SRTL distributions. Recall that a value of 0 for skewness is only a necessary condition for symmetry in a pdf, not a sufficient one. For a SRTL distribution with $\alpha = 2$ the skewness is close to 0 for all values of $\beta > 1\frac{1}{3}$ while for a SRP ($\alpha = 1$) distribution it increases in the interval $\beta \in [1, 10]$ from 0 to ≈ 2.30. From Fig. 7.3D we observe similarly that the minimum kurtosis value of a SRGTL distribution (with $\alpha \in (1, 2]$) is strictly larger than that of the minimum obtained in the SRP family (i.e. that of the uniform member with $\alpha = 1$ and $\beta = 1$ in (7.9)).

7.3.6 Inverse cumulative distribution function

Random samples from RGTL distributions may straightforwardly be generated utilizing the inverse cdf technique. From (7.8) we derive that $\{1 - F^{-1}(z|\alpha, \beta)\}$, $z \in [0, 1]$ is one of the roots of the quadratic equation in y

$$(\alpha - 1)y^2 - \alpha y + \sqrt[\beta]{1 - z} = 0. \tag{7.34}$$

Noting that (similarly to equation (7.18)) the symmetry axis associated with the LHS of the quadratic (7.34) has the value $\frac{1}{2}\alpha/(\alpha - 1)$ (Eq. (7.19)) which is strictly larger than 1 for $1 < \alpha \leq 2$, it follows that out of the two solutions of (7.34) only the solution

$$\frac{\alpha - \sqrt{\alpha^2 - 4(\alpha - 1)\sqrt[\beta]{1 - z}}}{2(\alpha - 1)}$$

can yield $\{1 - F^{-1}(z|\alpha, \beta)\} \in [0, 1]$. Analogously, it follows that for $0 \leq \alpha < 1$ only the second solution

$$\frac{\alpha + \sqrt{\alpha^2 - 4(\alpha - 1)\sqrt[\beta]{1 - z}}}{2(\alpha - 1)}$$

can result in $\{1 - F^{-1}(z|\alpha, \beta)\} \in [0, 1]$. Hence, the inverse cdf is given by

$$F^{-1}(z|\alpha, \beta) = \begin{cases} 1 - \frac{\alpha - \sqrt{\alpha^2 - 4(\alpha - 1)\sqrt[\beta]{1 - z}}}{2(\alpha - 1)}, & \text{for } 1 < \alpha \leq 2 \\ 1 - \sqrt[\beta]{1 - z}, & \text{for } \alpha = 1 \\ 1 - \frac{\alpha + \sqrt{\alpha^2 - 4(\alpha - 1)\sqrt[\beta]{1 - z}}}{2(\alpha - 1)}, & \text{for } 0 \leq \alpha < 1, \end{cases} \tag{7.35}$$

where the case $\alpha = 1$ follows from the cdf of a standard reflected power (SRP) distribution ($\alpha = 1$ in (7.8)). The inverse cdf is thus expressed explicitly in terms of elementary functions (which is not the case for the beta distribution (Eq. (4.1)) in Chapter 4). Hence, random sample realizations from a RGTL distribution are straightforward.

7.4 Maximum Likelihood Estimation of SRGTL Parameters

We have seen in Sec. 7.3 that the structure of the pdf and cdf and probabilistic properties of SRGTL distributions are appealingly direct. Moment expressions for SRGTL, unfortunately, are rather cumbersome requiring numerical search routines when one would like to estimate the SRGTL parameters α and β via the method of moments technique. Fortunately, the ML method does not present any intrinsic difficulties.

We shall discuss an approximate ML estimation procedure for a total of N observations grouped in m intervals $[x_{i-1}, x_i]$ with n_i observations each and interval mean values $\bar{x}_i$, where $x_0 \equiv 0$, $x_m \equiv 1$ and

$$N = \sum_{i=1}^{m} n_i.$$

The data described above may be summarized in an m-vector $\underline{x}$ whose elements are the interval mean values and an m-vector $\underline{n}$ containing the number of observations in each interval. The approximate ML estimation procedure below may easily be modified to an exact ML estimation procedure utilizing order statistics (as it was done for various TSP distributions in Chapters $3 - 6$), but here our approach is tailored to the format of the rather extensive US income distribution data to be presented in the detailed Table 7.1. The approximate ML procedure will assume that the probability mass is concentrated at the interval mean $\overline{x}_i$ of the subdividing intervals $[x_{i-1}, x_i]$. Utilizing (7.9), we have the likelihood $\mathcal{L}(\alpha, \beta | \underline{x}, \underline{n})$ to be proportional to

$$\beta^N \prod_{i=1}^{m} \left[\left\{ \alpha y_i - (\alpha - 1)y_i^2 \right\}^{\beta-1} \left\{ \alpha - 2(\alpha - 1)y_i \right\} \right]^{n_i} \qquad (7.36)$$

where

$$y_i = 1 - \overline{x}_i . \qquad (7.37)$$

As it is often the case, instead of maximizing $\mathcal{L}(\alpha, \beta | \underline{x}, \underline{n})$ we shall equivalently maximize the log-likelihood. Taking the logarithm of (7.36) and calculating the derivative with respect to β we obtain

$$\frac{N}{\beta} + \sum_{i=1}^{m} n_i Log \left\{ \alpha y_i - (\alpha - 1)y_i^2 \right\}. \qquad (7.38)$$

It thus follows from (7.38) that the estimator

$$\widehat{\beta} = N \left[\sum_{i=1}^{m} n_i Log \left\{ \frac{1}{\alpha y_i - (\alpha - 1)y_i^2} \right\} \right]^{-1} \qquad (7.39)$$

is the unique ML estimator of β for a given particular value of α. Taking the logarithm of (7.36) and calculating the derivative with respect to α, one obtains

$$(\beta - 1) \sum_{i=1}^{m} \frac{n_i \{1 - y_i\}}{\alpha - (\alpha - 1)y_i} + \sum_{i=1}^{m} \frac{n_i \{1 - 2y_i\}}{\alpha - 2(\alpha - 1)y_i}. \qquad (7.40)$$

Substituting (7.39) into (7.40) (namely utilizing $\widehat{\beta}$ instead of β and expressing $\widehat{\beta}$ in terms of α) the following function $\mathcal{G}(\alpha)$ emerges :

$$\mathcal{G}(\alpha) = \left[\frac{N}{\sum_{i=1}^{m} n_i Ln\left\{ \frac{1}{\alpha y_i - (\alpha - 1) y_i^2} \right\}} - 1 \right] \sum_{i=1}^{m} \frac{n_i \{1 - y_i\}}{\alpha - (\alpha - 1) y_i} + \quad (7.41)$$

$$\sum_{i=1}^{m} \frac{n_i \{1 - 2y_i\}}{\alpha - 2(\alpha - 1) y_i},$$

where y_i is given by (7.37) and the function is defined on a bounded range $0 \le \alpha \le 2$. As always the reader is encourages to verify the validity of (7.41). The ML estimator $\widehat{\alpha}$ now turns out to be either one of the roots of the equation $\mathcal{G}(\alpha) = 0$ or one of the boundary values: $\alpha = 0$ or $\alpha = 2$. The bounded domain of $\mathcal{G}(\alpha)$ allows for straightforward plotting of the function in standard spreadsheet software such as Microsoft EXCEL and subsequent determination of an approximate solution of the ML estimator $\widehat{\alpha}$. An illustrative example in the form of a graph will presented in the next section. Using the root finding algorithm GOALSEEK, available in Microsoft EXCEL (or a similar algorithm), and an approximate solution to the equation $\mathcal{G}(\alpha) = 0$ allows us to calculate $\widehat{\alpha}$ up to a desired level of accuracy. Finally, substitution of $\widehat{\alpha}$ into (7.39) yields the ML estimator $\widehat{\beta}$ corresponding to the ML estimator $\widehat{\alpha}$. The ML procedure above will be demonstrated in the next section using the U.S. 2001 household income data.

7.5 Fitting 2001 US Household Income Distribution Data

In the lead article in a recent issue of the *Journal of the American Statistical Association* (2002, Vol. 97, pp. 663-673) co-authored by Barsky *et al.*, an illuminating and comprehensive analysis of the African-American and Caucasian (Non-Hispanic) wealth gap was presented based on a longitudinal survey of some 6000 households over the period of 1968-1992. The authors argue, *interalia*, that a parametric estimation of the wealth-earning relationship by race is not an appropriate approach. Their main contention is that the wealth-earning relationship is non-linear of a *unknown* functional form which is difficult to parameterize and parametric estimation may thus likely yield unreliable estimates. The authors also provide an extensive and up-to-date bibliography of relevant books and articles up to and including 2001. They note that the racial *wealth* gap far exceeds the racial *income* gap especially at the higher wealth ranges, suggesting that the racial wealth gap is too large to be satisfactorily explained by income gap alone. On the other hand, they conclude that the role of earnings differences in explaining this gap is the most prominent at the lower tails of the wealth distribution and decreases dramatically at higher wealth levels. In fact, their results indicate that differences in the household earnings account for all of the racial wealth difference in the first lower quartile of the wealth distribution. Interested readers are also referred to Couch and Daly (2000) and O'Neill *et al.* (2002) who study the related topic of the racial wage gap in the U.S.A.

Our approach to this problem is somewhat different. We attempt to use the distribution developed in the previous sections and to fit − for the more recent household income data in the US for the year 2001[1] classified according to Caucasian (Non-Hispanic), African-American and Hispanic populations − a RGTL distribution using the above described ML procedure and draw some tentative conclusions about the racial income gap based on this data. Parametric estimation of income data has been quite common for almost 100 years and a wide variety of distributions have been proposed (see Kleiber and Kotz (2003) for an extensive bibliography). RGTL distributions (which are not discussed in Kleiber and Kotz (2003)) allow for a strictly positive density value at its *lower bound*, which is observed in a non-parametric kernel density estimate of the 1989 income data (see Fig. 2 in Barsky *et al.* (2002) p. 668). Fortunately, the new distribution that we are proposing seems to be appropriate for the US 2001 household income data, especially for that of the African-American sub population. We emphasize that the main purpose of the numerical analysis below is to illustrate fitting techniques for a RGTL distribution and properties of its parameters. The numerical analysis herein sheds light on the current state of affairs only and a further study is recommended. The analysis below in no way yields a conclusive answer to the problem of racial income gaps in the USA (nor that of racial wealth and racial wage gaps). Table 7.1 contains income distribution data for households in the year 2001 for the three different ethnic groups: Caucasian (Non-Hispanic), African-American and Hispanic in the U.S.A. Actually, the mean incomes in each bracket already tell part of the story (at least qualitatively).

---

[1]Source: U.S. Census Bureau, Current Population Survey, March 2002

Table 7.1 Income distribution for households in the U.S.A. for the year 2001
(Source: U.S. Census Bureau, Current Population Survey, March 2002.)
Numbers are in thousands households as of March of the following year.

Income of Household	Caucasion (non-Hispanic) Number	Caucasion (non-Hispanic) Mean Income	African American Number	African American Mean Income	Hispanic Number	Hispanic Mean Income
Under $2,500..............	1,443	$168	520	$439	273	$426
$2,500 to $4,999..........	773	$3,808	388	$3,842	137	$3,767
$5,000 to $7,499..........	2,141	$6,450	698	$6,359	304	$6,387
$7,500 to $9,999..........	2,561	$8,749	756	$8,658	404	$8,663
$10,000 to $12,499........	3,142	$11,220	621	$11,173	458	$11,214
$12,500 to $14,999........	2,946	$13,615	543	$13,672	411	$13,659
$15,000 to $17,499........	3,167	$16,091	660	$16,089	553	$15,993
$17,500 to $19,999........	2,803	$18,660	479	$18,655	418	$18,579
$20,000 to $22,499........	3,099	$21,082	610	$21,094	490	$21,005
$22,500 to $24,999........	2,697	$23,706	447	$23,682	373	$23,691
$25,000 to $27,499........	3,055	$26,064	570	$26,061	477	$26,011
$27,500 to $29,999........	2,446	$28,673	464	$28,544	330	$28,617
$30,000 to $32,499........	3,277	$31,059	492	$31,040	479	$30,998
$32,500 to $34,999........	2,330	$33,679	375	$33,655	335	$33,601
$35,000 to $37,499........	2,950	$36,045	437	$35,944	412	$36,082
$37,500 to $39,999........	2,114	$38,713	310	$38,626	249	$38,641
$40,000 to $42,499........	2,846	$41,052	434	$41,004	424	$40,938
$42,500 to $44,999........	1,924	$43,679	260	$43,693	231	$43,668
$45,000 to $47,499........	2,236	$46,058	289	$45,908	291	$46,044
$47,500 to $49,999........	1,966	$48,709	256	$48,655	205	$48,607
$50,000 to $52,499........	2,403	$51,042	350	$50,924	247	$51,021
$52,500 to $54,999........	1,730	$53,670	210	$53,553	153	$53,725
$55,000 to $57,499........	2,014	$56,127	249	$55,972	224	$55,992
$57,500 to $59,999........	1,528	$58,650	177	$58,680	177	$58,764
$60,000 to $62,499........	2,047	$61,053	248	$60,979	219	$61,106
$62,500 to $64,999........	1,417	$63,719	162	$63,761	141	$63,801
$65,000 to $67,499........	1,710	$66,048	175	$65,990	157	$66,018
$67,500 to $69,999........	1,325	$68,677	150	$68,705	124	$68,734
$70,000 to $72,499........	1,622	$71,067	190	$71,090	159	$71,112
$72,500 to $74,999........	1,248	$73,707	142	$73,589	128	$73,711
$75,000 to $77,499........	1,608	$75,981	133	$75,974	132	$75,860
$77,500 to $79,999........	1,073	$78,662	100	$78,693	72	$78,726
$80,000 to $82,499........	1,380	$81,051	100	$80,950	125	$80,976
$82,500 to $84,999........	993	$83,688	90	$83,584	90	$83,708
$85,000 to $87,499........	1,144	$86,057	103	$85,984	76	$85,830
$87,500 to $89,999........	803	$88,696	86	$88,754	55	$88,636
$90,000 to $92,499........	985	$91,051	78	$91,103	83	$90,997
$92,500 to $94,999........	701	$93,658	83	$93,666	41	$93,579
$95,000 to $97,499........	915	$96,071	71	$95,901	65	$95,999
$97,500 to $99,999........	712	$98,682	65	$98,639	48	$98,811
$100,000 to $149,999.......	8,374	$119,083	554	$117,549	515	$119,016
$150,000 to $199,999.......	2,689	$169,312	115	$172,222	113	$164,692
$200,000 to $249,999.......	993	$219,285	29	$218,672	43	$221,737
$250,000 and above.........	1,345	$462,675	46	$433,097	59	$474,843
Total	90,682	$60,512	13,315	$39,248	10,499	$44,383

The ML procedure discussed in Sec. 7.4 will be used to fit RGTL distributions for incomes of these three groups. Only the data up to \$250,000 can be used from Table 7.1 since the U.S. Census Bureau does not provide the maximum observed income in their statistics. Of the total number of U.S. households surveyed 98.58%, 99.44% and 99.65% have had in 2001 household incomes of value less than \$250,000 for the Caucasian (Non-Hispanic), Hispanic and African-American groups, respectively. Figure 7.4 displays a graph of the function $\mathcal{G}(\alpha)$ (Eq. (7.41)) for the income data of Caucasian (Non-Hispanic) Americans presented in Table 7.1. From Fig. 7.4 we observe an approximate root of the equation $\mathcal{G}(\alpha) = 0$ to be the value $\alpha^* \approx 1.70$. Since, $\mathcal{G}(\alpha) > 0$ (< 0) for $0 \leq \alpha < \alpha^*$ ($\alpha^* \leq \alpha \leq 2$) it follows that $\widehat{\alpha} = \alpha^*$ is the unique ML estimator for α based on the likelihood (7.36). Using GOALSEEK (a standard root-finding algorithm in the Microsoft EXCEL software) with an accuracy of $1 \cdot 10^{-6}$ and utilizing the approximate solution $\alpha^* \approx 1.70$ we arrive at the more precise estimator $\widehat{\alpha} = \alpha^* = 1.679$. The unique ML estimator $\widehat{\beta} = 6.767$ follows from substituting $\widehat{\alpha} = 1.679$ into (7.39).

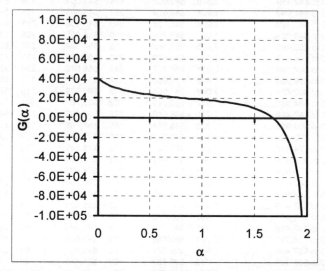

Fig. 7.4 A graph of the function $\mathcal{G}(\alpha)$ (Eq. (7.41)) for the income data of Caucasian (Non-Hispanic) Americans presented in Table 7.1.

Figure 7.5 below plots both the empirical cdf and its fitted RGTL counterpart (Eqs. (7.2) and (7.4)) with $a = \$0, b = \$250,000, \alpha = 1.679$ and $\beta = 6.767$. Differences between the empirical and fitted cdf's can be observed in Fig. 7.5A. The well-known and widely used Kolmogorov-Smirnov statistic D, whose value is the maximum of the observed difference between the empirical and fitted cdf's (see, e.g., DeGroot (1991)) in Fig. 7.5A equals 8.60%. Hence, with 43 degrees of freedom (Table 7.1 has 43 rows up to $250,000) the Kolmogorov-Smirnov test accepts the fitted RGTL distribution at the 10% $(D_{0.10} \approx 0.182)$, 5% $(D_{0.05} \approx 0.203)$ and even at 1% $(D_{0.01} \approx 0.243)$ levels. Please note however a quite substantial deviation of empirical pdf values from the fitted ones in the income range from \$10,000-\$30,000 and in the vicinity of \$100,000. In the first (second) case the fitted pdf underestimates (over estimates) the empirical pdf.

Table 7.2 provides the unique ML estimators for $\widehat{\alpha}$ and $\widehat{\beta}$ (obtained using the procedure described in Sec. 7.4) for the Caucasian (Non-Hispanic), African-American and Hispanic income data presented in Table 7.1. Note that values of $\widehat{\beta}$ are very similar for the Hispanic and African American incomes and the Caucasian (Non-Hispanic) value is much lower. Clearly we are dealing with Case 3 in the Sec. 7.3.3. devoted to mode analysis.

Table 7.2 Maximum likelihood estimators for the parameters $\widehat{\alpha}$ and $\widehat{\beta}$ of RGTL distributions for the income data in Table 7.1 (up to \$250,000)

	$\widehat{\alpha}$	$\widehat{\beta}$
$Caucasian\,(Non\text{-}Hispanic)$	1.679	6.767
$Hispanic$	1.685	10.306
$African\,American$	1.613	10.629

Similarly, Fig. 7.6A (Fig. 7.6B) plots the empirical and fitted RGTL pdf with ML estimators $\widehat{\alpha} = 1.613$, $\widehat{\beta} = 10.629$ ($\widehat{\alpha} = 1.685$ and $\widehat{\beta} = 10.306$) for the African-American (Hispanic) income data presented in Table 7.1. The Kolmogorov-Smirnov Statistic D for the African-American (Hispanic) income data equals 6.01% (8.09%) which is slightly smaller than that of the Caucasian (Non-Hispanic) income data (indicating a better fit). Hence the Kolmogorov-Smirnov test accepts the ML fitted RGTL

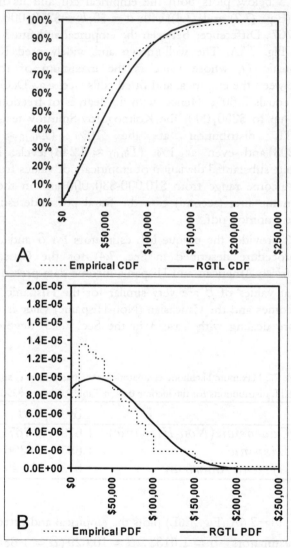

Fig. 7.5 Empirical and a ML fitted RGTL cdf (A) and pdf (B) $(\widehat{\alpha} = 1.643$ and $\widehat{\beta} = 6.179)$ of the Caucasian (Non-Hispanic) income data presented in Table 7.1.

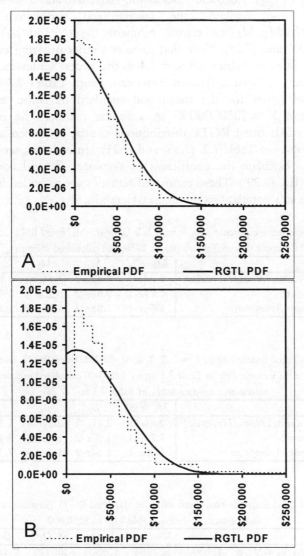

Fig. 7.6 Empirical and ML fitted RGTL pdf's for the income data in Table 7.1;
A: African-American $(\widehat{\alpha} = 1.613, \widehat{\beta} = 10.629)$; B: Hispanic $(\widehat{\alpha} = 1.685, \widehat{\beta} = 10.306)$.

distributions in Figs. 7.6A and 7.6B at the 10%, 5% and 1% levels. Table 7.3 (Table 7.4) contains the standardized cumulative moments $M_0 = \mu'_1, M_1, M_2, M_3$ (the central moments μ_2, μ_3 and μ_4) calculated utilizing (7.33) and (7.28). Note that there is a strict ordering column-wise for all the values in Tables 7.3 and 7.4 in the order: Caucasian American (non-Hispanic), Hispanic, African-American. From Tables 7.3 and 7.4 we can calculate values for the mean and standard deviation utilizing the transformation $Y = \$250,000X$. In a similar manner, the median and mode of the ML fitted RGTL distributions can be evaluated utilizing the parameter values in Table 7.2, (7.35) and (7.21). In addition, we may utilize Table 7.4 to calculate the coefficient of skewness β_1 and coefficient of kurtosis β_2 (Eq. (7.29)). These estimated statistics are provided in Table 7.5 for the three sub-populations under consideration.

Table 7.3. Cumulative moments M_k, $k = 0, 1, 2, 3$, of the ML fitted RGTL distributions for the income data in Table 1 (up to \$250,000) calculated utilizing (7.33).

	$M_0 = \mu'_1$	M_1	M_2	M_3
Caucasian (Non-Hispanic)	2.34e-1	3.97e-2	1.11e-2	3.80e-3
Hispanic	1.77e-1	2.38e-2	5.26e-3	1.51e-3
African American	1.59e-1	1.98e-2	4.17e-3	1.14e-3

Table 7.4 Central moments μ_k, $k = 1, 2, 3, 4$, of the ML fitted RGTL distributions for the income data in Table 7.1 up to \$250,000 calculated utilizing cumulative moments M_k in Table 7.3 and (7.28).

	$M_0 = \mu'_1$	μ_2	μ_3	μ_4
Caucasian (Non-Hispanic)	2.34e-1	2.47e-2	2.54e-3	1.75e-3
Hispanic	1.77e-1	1.60e-2	1.66e-3	8.40e-4
African American	1.59e-1	1.44e-2	1.60e-3	7.31e-4

Table 7.5 Statistics associated with the ML fitted RGTL distributions for the income data in Table 1 up to \$250,000.

	Mean	*Median*	*Mode*	*St. Dev*	β_1	β_2
Caucasian (Non-Hisp.	\$58393	\$52534	\$28306	\$39326	0.424	2.858
Hispanic	\$44316	\$38606	\$11851	\$31710	0.660	3.248
African American	\$39786	\$33599	\$0	\$30002	0.858	3.522

A similar ordering to the one observed in Tables 7.3 and 7.4 is presented throughout Table 7.5. Note that the difference in the point estimates in Table 7.5 between the Caucasian (Non-Hispanic) population and the African-American Population is approximately $18607 or more and those associated with the Hispanic population differ by the amount $13928 or more. The latter observation is amplified somewhat in Table 7.5 by the fact that the fitted mean income for the Caucasian (Non-Hispanic) population overestimates the empirical mean (for incomes up to $250, 000) by $3936 whereas the fitted mean income for the African-American (Hispanic) population is overestimated by only $1898 ($2357). Perhaps the most notable feature is the modal income value of $0 for the ML fitted RGTL distribution for the African-American population while the modal income value for the Caucasian (Non-Hispanic) and Hispanic population have a value substantially larger than zero (and moreover the mode for the Caucasian (Non-Hispanic) population is more than twice that of Hispanics). A similar observation can be made by comparing the RGTL distributions in Figs. 7.5B, 7.6A and 7.6B. Finally, from Table 7.2 and (7.5) we may evaluate the density values at the lower bound, i.e. $f(0|0, \$250, 000, \widehat{\alpha}, \widehat{\beta})$, presented in Table 7.6. Hence, in comparison with Americans of Caucasian origin, African-Americans appear to be approximately 1.9 times as likely and Hispanics 1.5 times as likely to have negligible income in the year 2001. It is the fact that our ML fitted RGTL pdf's may take a positive value at its lower bound while allowing a strictly positive mode, that allows us to reach such an indicative conclusion.

Table 7.6. Density values at the lower bound of the ML fitted RGTL distributions for the income data in Table 1 (up to $250,000).

	$f(0|0, \$250, 000, \widehat{\alpha}, \widehat{\beta})$
Caucasian (Non-Hispanic)	8.68e-6
Hispanic	1.30e-5
African American	1.65e-5

A further analysis of the obtained results sheds an additional light on the differences in the income distributions for the 3 ethnic groups under examination. In Fig. 7.7A, we utilize the ML fitted RGTL income distributions by plotting the percentiles of the African-American and Hispanic income distributions against the Caucasian (Non-Hispanic) data using (7.8), (7.35) and the corresponding ML values for $\widehat{\alpha}$ and $\widehat{\beta}$ in Table

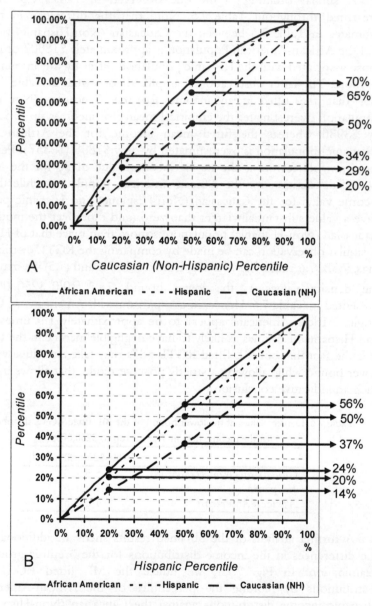

Fig. 7.7 Stochastic dominance analysis by ethnicity for
the income data presented in Table 7.1 utilizing the ML fitted RGTL cdf's.

7.2. For example, from Fig. 7.7A, we observe that approximately 70% (65%) of the African-American (Hispanic) population have less income than the median (50%)of the Caucasian (Non-Hispanic) income distribution. Similar comparisons can be made for other percentiles of the Caucasian (Non-Hispanic) income distribution utilizing the same figure. For example, 34% (29%) of the African-American (Hispanic) population earn less than what the lowest 20% of the Caucasian (Non-Hispanic) population earn (i.e. the 20% percentile of the Caucasian (Non-Hispanic) income distribution). Note that the solid curves in Fig. 7.7A involving the African-American and Hispanic income distributions are located completely above the unit diagonal implying the stochastic dominance of Caucasian (Non-Hispanic) income over that of the African-American or Hispanic ones. The latter can be directly concluded from the ML values for $\widehat{\alpha}$ and $\widehat{\beta}$ in Table 7.2 and Eq. (7.14) for the African-American and Caucasian (Non-Hispanic) comparison but not for the Hispanic and Caucasian (Non-Hispanic) comparison. This shows that the implication arrow in Eq. (7.14) cannot in general be reversed.

In a similar manner, Fig. 7.7B utilizes the ML fitted RGTL income distributions by plotting the percentiles of the African-American and Caucasian (Non-Hispanic) income distributions against those of the Hispanic one. For example, from Fig. 7.7B, we observe that approximately 56% (37%) of the African-American (Caucasian Non-Hispanic) population have less income than the median (50%) of the Hispanic income distribution. Additional illuminating comparisons can be deduced from Figs. 7.7A and 7.7B. We now conclude from Fig. 7.7B that Hispanic income in 2001 stochastically dominates the African-American one. The latter conclusion also follows immediately from the corresponding ML estimator values for $\widehat{\alpha}$ and $\widehat{\beta}$ in Table 7.2 and Eq. (7.14). However, we can conclude, once again, only by observation in Fig. 7.7B that Hispanic income is stochastically dominated by Caucasian (Non-Hispanic) income (since the line associated with the Caucasian (Non-Hispanic) is now completely below the unit diagonal). As above, this conclusion cannot be directly obtained from the corresponding ML values for $\widehat{\alpha}$ and $\widehat{\beta}$ in Table 7.2 and Eq. (7.14).

Summarizing, Table 7.2 and Eq. (7.14) alone imply that the chances of a Caucasian (Non-Hispanic) or a Hispanic American earning more than a specified amount (within the range from $0 to $250,000) are higher than those for an African-American. In addition, the analysis in Fig. 7.7 allows us to conclude that the chances of a Caucasian (Non-Hispanic) earning more than a specified amount (within the same range from $0 to $250, 000) are higher than those of a Hispanic. Moreover, Fig. 7.7 and Table 7.5 demonstrate that although substantial advances have reportedly been made in reducing the income distribution gap amongst these three subpopulations in the U.S. during the last 20 years or so (see, e.g., Couch and Daly (2000)), these differences still exist and are quite noticeable. The readers are encourages to repeat this analysis for the U.S. income data of 2003 (or the latest year for which now data is available). It is also of interest to investigate the income gaps in other countries with the population belonging to several income groups.

A perceptive reader will undoubtedly notice that the ML fitted RGTL distributions consistently underestimate the empirical pdf in Figs 7.5B, 7.6A and 7.6B in the lower income ranges (perhaps due to the nature of record keeping and data collection in these ranges). However, the conclusions above related to the income gaps amongst the three subpopulations (which utilized ML fitted RGTL distributions) would seem not to be affected or only augmented due to this fitting phenomenon.

Chapter 8

A Generalized Framework for
Two-Sided Distributions

In this chapter we shall present a generalized structure of the Two-Sided Power distribution discussed in Chapters 3 and 4. These distributions will be referred to belong to the Two-Sided (T-S) family of distributions. Each sub-family in the Standard TS (ST-S) framework is characterized by a *generating density* with a bounded support $[0, 1]$. Standard T-S members can be transformed to T-S ones (with an arbitrary finite support $[a, b]$) by means of a linear transformation. Some examples of ST-S distributions will be presented using various generating densities and the properties of the *Two Sided Slope* (T-SS) family (using the linear slope distribution as its generating density) will be further examined. Our readers are encouraged to derive properties of other TS families described in this chapter. Our reasons for a further investigation of the T-SS family are two-fold. Firstly, the T-SS density is a natural *linear* extension of the triangular distribution discussed in Chapter 1 and secondly, the slope distribution serves as a link between the Generalized trapezoidal (Chapter 5), the Generalized TSP and the Uneven TSP (both discussed in Chapter 6) and finally, the Reflected Generalized Topp and Leone distributions (described in Chapter 7). It may be appropriate at this time to review the connections amongst these distributions as presented in *Appendix A*. We also remind our reader again that the present chapter can be studied directly after Chapter 3 (dealing with STSP distributions) without loss of continuity.

8.1 Standard Two-Sided Families of Distributions

The Standard Two-Sided Power (STSP) distributions introduced in Chapter 3 (Eq. (3.11)) can be motivated from at least two different aspects. The initial motivation is to extend the triangular distribution (see, Chapter 3). A more encompassing approach to STSP distributions is the realization that the STSP density (3.11) can be viewed as a particular case of the *general* Standard Two-Sided (STS) continuous family with support $[0, 1]$ given by the density

$$g\{x|\theta, p(\,\cdot\,|\Psi)\} = \begin{cases} p(\frac{x}{\theta}|\Psi), & \text{for } 0 < x \leq \theta \\ p(\frac{1-x}{1-\theta}|\Psi), & \text{for } \theta < x < 1, \end{cases} \tag{8.1}$$

where $p(\,\cdot\,|\Psi)$ is an appropriately selected continuous pdf defined on $[0, 1]$ with parameter(s) Ψ, which may in principle be vector-valued. The density $p(\,\cdot\,|\Psi)$ will be referred to as the *generating density* of the resulting STS family of distributions and the parameter θ is termed the reflection parameter (an alternative designation could be the *hinge* or *threshold* parameter). We note that the concepts of generating density and hinge were introduced in the book on several occasions.

The simplest linear choice for the generating density

$$p(y) = 2y, \quad 0 \leq y \leq 1, \tag{8.2}$$

results in the triangular distribution (see Fig. 8.1A). A more general form of $p(y|\,\cdot\,)$ is

$$p(y|n) = ny^{n-1}, \ 0 \leq y \leq 1, n > 0, \tag{8.3}$$

which generates via (8.1) the STSP distribution (3.11) (see Fig. 8.1B) extending the triangular distribution. The density of a normalized truncated exponential distribution, i.e.,

$$p(y|\lambda) = exp(-\lambda y)/\{1 - e^{-\lambda}\}, \ 0 \leq y \leq 1, \lambda > 0,$$

generates the two sided truncated exponential distribution (see Fig. 8.1C). The density

$$p(y|\alpha) = \alpha + 2(1-\alpha)y, \ 0 \leq y \leq 1, \ 0 \leq \alpha \leq 2, \tag{8.4}$$

which was encountered earlier in Chapter 7 and was called the *slope distribution*, results in the Standard Two-Sided Slope (ST-SS) distribution (see, Fig. 8.1D). The pdf (8.4) which was used in connection with the generalization of the Topp and Leone distribution (GTL) in Chapter 7 (see, Eq. (7.1)) and — to the best of our knowledge — has not been as yet discussed in the relevant periodical literature.

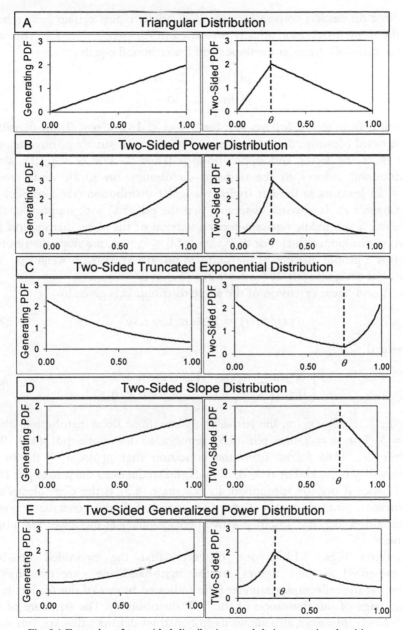

Fig. 8.1 Examples of two-sided distributions and their generating densities.

For the readers convenience we shall briefly repeat certain facts related to the slope distribution already presented in Chapter 7, Sec. 7.1. The ratio of the pdf (8.4) values at the upper and lower bound equals

$$\frac{p(1|\alpha)}{p(0|\alpha)} = \frac{2-\alpha}{\alpha}$$

and thus from (8.4) it follows that for $0 < \alpha < 1$ $(1 < \alpha < 2)$ the density is an upward (downward) sloping linear function with strictly positive density values at the lower and upper bounds 0 and 1. For $\alpha = 1$, the slope distribution reduces to the uniform distribution on $[0, 1]$, while $\alpha = 0$ $(\alpha = 2)$ leads us to the left (right) triangular distribution (see also Sec. 7.1 of Chapter 7). It is worth observing that the pdf (5.6) with support $[b, c]$ in Chapter 5 is a mildly reparameterized version of the slope distribution pdf (8.4) with support $[0, 1]$. For the values of $0 \leq \alpha \leq 1$, the slope distribution (8.4) is a proper mixture of the uniform pdf on $[0, 1]$ (with weight α) and the right triangular pdf $2y$ (with weight $(1 - \alpha)$).

A non-linear extension of the slope distribution is given by

$$p(y|\alpha, n) = \alpha + n(1 - \alpha)y^{n-1} \tag{8.5}$$

where $0 \leq y \leq 1$,

$$\begin{cases} 0 \leq \alpha \leq n/(n-1), & \text{for } n > 1 \\ 0 \leq \alpha \leq 1, & \text{for } 0 < n \leq 1. \end{cases} \tag{8.6}$$

Figure 8.1E depicts (on the left side) a generalized slope distribution where $n = 3$. For $n = 2$, the pdf (8.5) reduces to the slope pdf (8.4). The restrictions (8.6) follow from the condition that $p(y|\alpha, n) \geq 0$, for all values of $y \in [0, 1]$. For $\alpha = 0$, the pdf (8.5) reduces to the power pdf (8.3) and hence it may be appropriate to designate (8.5) as the *Generalized Power distribution*. Similarly to the slope distribution (8.4) the generalized power distribution (8.5) has strictly positive density values at the lower and upper bounds.

From Figs. 8.1A-E one observes that the two-sided densities − presented in these figures on the right-hand-side − are non-smooth curves at the reflection parameter θ (as indicated by one of our readers in an early stage of our derivation of the TSP distribution). The structure of the STS family, however, also allows us to construct densities that are "smooth" at the threshold parameter θ. Indeed, by setting

$$Y^n = X, \ n > 0,$$

where $X \sim p(\cdot \mid \alpha)$ given by Eq. (8.4) we obtain the two-parameter density for Y:

$$p(y \mid \alpha, n) = ny^{n-1}\{\alpha - 2(\alpha - 1)y^n\}, \ n > 0, \ 0 \le \alpha \le 2, \quad (8.7)$$

with support $[0, 1]$. Next, setting

$$\partial p(y \mid \alpha, n)/\partial y \big|_{y=1} = 0$$

and solving for α we have

$$\alpha = \frac{4n - 2}{3n - 1} \quad (8.8)$$

for $n > 0$. Utilizing the parameter restrictions $0 \le \alpha \le 2, n > 0$ we obtain from (8.8) that $n \ge 1/2$. Hence, substituting (8.8) into the pdf (8.7), we arrive at the one-parameter pdf:

$$p(y \mid \alpha, n) = ny^{n-1}\{\frac{4n - 2}{3n - 1} - \frac{2n - 2}{3n - 1}y^n\}, \ 0 \le y \le 1, n \ge 1/2, \quad (8.9)$$

to be referred to as the *ogive distribution*. The pdf (8.9) should not be confused with the original designation "ogive curve" due to Galton in 1875 which is occasionally used in older books on statistical methodology for S-shaped curves (see, e.g., Stigler (1986)). The symmetric $(\theta = \frac{1}{2})$ Standard Two-Sided Ogive (STSO) distribution (constructed utilizing (8.9) and the basic structure (8.1)) is depicted in Fig. 8.2.

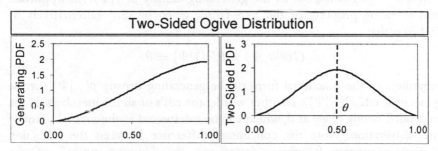

Fig. 8.2 A symmetric two-sided ogive distribution and its generating density (8.9).

Note that STSO distributions are smooth at the reflection parameter θ (unlike most of the other members in the STS family) and may thus provide a *smooth* alternative to the beta distribution. For $n = \frac{1}{2}$ and $n = 1$, the pdf (8.9) reduces to a uniform distribution on $[0, 1]$. For $\frac{1}{2} < n < 1$, $\theta \in (0, 1)$ ($\theta = 0$ or $\theta = 1$) the STSO distribution becomes U-shaped (J-Shaped) and for $n > 1$ the distribution is uni-modal with the mode at θ.

Summarizing, the examples presented in Figs. 8.1 and 8.2 portray two strictly increasing convex generating densities (Fig. 8.1B and E), a strictly decreasing convex one (Fig. 8.1C), two increasing linear generating densities (Fig. 8.1A and D, with the second one taking a positive density value at the origin) and a generating density possessing an inflection point (Fig. 8.2). Evidently, a large variety of bounded continuous distributions can be constructed using the outlined procedure. Two-Sided distributions with a support $[a, b]$ can be obtained from their STS counterparts by means of a simple linear transformation. Note that our model is different in its structure (although similar in the spirit) from the double Weibull distribution originally introduced in Balakrishnan and Kocherlakota (1985) some 20 years ago.

Continuing our investigation of the general structure of STS families of distributions we obtain for the cdf associated with the pdf (8.1) the expression:

$$G\{x|\theta, P(\cdot \ \Psi)\} = \qquad\qquad\qquad (8.10)$$
$$\begin{cases} \theta P(\frac{x}{\theta}|\Psi), & \text{for } 0 < x < \theta \\ 1 - (1 - \theta)P(\frac{1-x}{1-\theta}|\Psi), & \text{for } \theta \leq x < 1, \end{cases}$$

where $P(\cdot \ |\Psi)$ is the cdf of the generating density $p(\cdot \ |\Psi)$. An important and revealing property of the STS family given by (8.1) (alternatively by (8.10)) is that

$$G(\theta|\theta, \Psi) = \theta P(1|\Psi) = \theta$$

regardless of the functional form of the generating density $p(\cdot \ |\Psi)$ (or the generating cdf $P(\cdot \ |\Psi)$). In other words, the cdf's of all the members of the two-sided family hinge at θ, which can be interpreted as the pivotal point of the distribution. Note the conceptual difference between the reflection (hinge) parameter θ which determines the "turning point" of the distribution under consideration and the parameters included in Ψ which control the form of the two sides of the distribution (to the left and the right of θ).

If $X \sim g\{ \cdot \,|\theta, p(\cdot \,|\Psi)\}$ (Eq. (8.1)) and $Y \sim p(\cdot \,|\Psi)$, the following relationship between the moments around zero of X and Y can straightforwardly be derived by induction:

$$E[X^k|\theta, \Psi] = \theta^{k+1} E[Y^k|\Psi] +$$
$$\sum_{i=0}^{k} \binom{k}{i} (-1)^i (1-\theta)^{i+1} E[Y^i|\Psi].$$

(Recall Y is distributed in accordance with the generating density and X is the corresponding STS random variable.) From (8.14), utilizing modern computational facilities (especially for large value of k), moments of two-sided distributions can be calculated at least in the case when closed form expressions for the moments of the generating density $p(\cdot \,|\Psi)$ are available. In particular, we have for the first two moments

$$E[X|\theta, \Psi] = (2\theta - 1)E[Y|\Psi] + (1-\theta) \tag{8.11}$$

and

$$Var(X|\theta, \Psi) = \{1 - 3\theta + 3\theta^2\} Var(Y|\Psi) + \tag{8.12}$$
$$\theta(1-\theta)\{E[Y|\Psi] - 1\}^2.$$

Note that for $\theta = 1, 0$ and $\frac{1}{2}$ it follows from (8.11) that

$$\begin{cases} E[X|1, \Psi] = E[Y|\Psi], & \text{for } \theta = 1, \\ E[X|0, \Psi] = 1 - E[Y|\Psi], & \text{for } \theta = 0, \\ E[X|\frac{1}{2}, \Psi] = \frac{1}{2}, & \text{for } \theta = \frac{1}{2}, \end{cases} \tag{8.13}$$

while from (8.12) we obtain

$$Var(X|1, \Psi) = Var(X|0, \Psi) = Var(Y|\Psi). \tag{8.14}$$

The structure of the density (8.1) implies for $\theta = 1$ that

$$g\{x|\theta, \ p(\cdot \,|\Psi)\} = p(x|\Psi).$$

Hence, the relations for $\theta = 1$ in (8.13) and (8.14) are trivially verified. The second result $(\theta = 0)$ in (8.13) and (8.14) follows from the observation that $g\{x|0, p(\cdot \,|\Psi)\}$ represents a mirror image of the generating density, i.e.

$p(1 - x|\Psi)$. The third result in (8.13) holds regardless of the form of $p(x|\Psi)$ due to the symmetry of $g\{x|\frac{1}{2}, p(\cdot|\Psi)\}$ around $\theta = \frac{1}{2}$.

In the remainder of this chapter, we shall further investigate the properties of the Two-Sided Slope (T-SS) distribution (see, Fig. 8.1D) within the framework above. In our opinion, a T-SS distribution it is a natural linear extension of the triangular distribution (discussed in Chapter 1). Moreover, the slope distribution has been shown to serve as a *linking* distribution amongst various bounded distributions discussed in Chapter 5 (the generalized trapezoidal distribution), Chapter 6 (the GTSP and UTSP distributions) and in Chapter 7 (the RGTL distribution). The reader is encouraged to conduct similar investigations for the ogive distribution (8.9), the generalized power distribution (8.5) and the truncated exponential distribution (8.4) together with their two-sided counterparts (which all are some of the members of the family given by the density (8.1)).

8.2 The Two-Sided (Linear) Slope Distribution

From the general form of the pdf of an STS distribution (8.1) and the pdf

$$p(y|\alpha) = \alpha + 2(1 - \alpha)y, \ 0 \le y \le 1, 0 \le \alpha \le 2$$

of a slope distribution (8.4) we have for the Standard Two-Sided Slope (ST-SS) distribution the pdf

$$g(x|\theta, \alpha) = \begin{cases} \alpha + 2(1 - \alpha)\frac{x}{\theta}, & \text{for } 0 < x \le \theta \\ \alpha + 2(1 - \alpha)\frac{1-x}{1-\theta}, & \text{for } \theta < x < 1, \end{cases} \tag{8.15}$$

where $0 \le x \le 1$, $0 \le \alpha \le 2$, $0 \le \theta \le 1$. Figure 8.3 plots ST-SS pdf's for increasing values of α. This includes a Triangular ($\alpha = 0$), unimodal ST-SS ($\alpha = 0.5$), Uniform ($\alpha = 1$), U-shaped TSS ($\alpha = 1.5$) and a *reverse triangular* pdf ($\alpha = 2$). We are using the designation *reverse* triangular since the pdf in 8.3E is a reflection through a *horizontal* line of the distribution in Fig. 8.3A. The designation *reflected* has been utilized elsewhere in the book to indicate reflection through a *vertical* line (see, e.g. the RGTL distribution in Chapter 7). Note that for unimodal (U-Shaped) ST-SS distributions — $0 \le \alpha \le 1$ $(1 \le \alpha \le 2)$ — the mode (the anti-mode) is attained at θ corresponding to a density value $(2 - \alpha)$ and the anti-mode

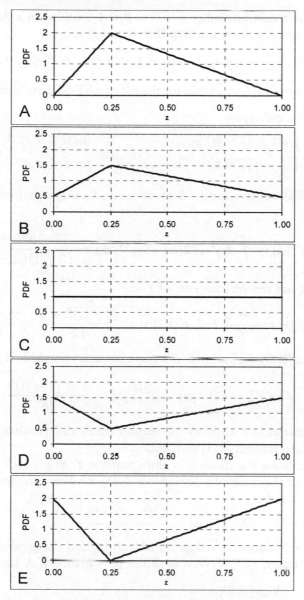

Fig. 8.3 Standard two-sided slope (ST-SS) pdf's (8.15)
A: $\alpha = 0$ (triangular), B: $\alpha = 0.5$ (unimodal ST-SS), C: $\alpha = 1$ (uniform),
D: $\alpha = 1.5$ (U-shaped ST-SS), E: $\alpha = 2$ (reverse triangular).

233

(the mode) is attained at either $\theta = 0$ or $\theta = 1$ resulting in a density value α. For the slope distribution with pdf (8.4) the cdf is

$$P(y|\alpha) = \alpha y + (1 - \alpha)y^2, 0 \leq y \leq 1,\ 0 \leq \alpha \leq 2,$$

and hence from (8.10) we obtain for the cdf of a ST-SS distribution

$$G(x|\theta, \alpha) = \tag{8.16}$$
$$\begin{cases} \alpha x + (1 - \alpha)\frac{x^2}{\theta}, & \text{for } 0 < x < \theta \\ 1 - \alpha(1 - x) - (1 - \alpha)\frac{(1-x)^2}{(1-\theta)}, & \text{for } \theta \leq x < 1. \end{cases}$$

From (8.4) we have for the k-th moment of a slope random variable Y to be:

$$E[Y^k|\alpha] = \int_0^1 \alpha y^k - 2(\alpha - 1)y^{k+1} dy = \frac{(2 - \alpha)k + 2}{(k + 1)(k + 2)}$$

and the first and second moments of Y are

$$E[Y|\alpha] = \frac{4 - \alpha}{6} \text{ and } E[Y^2|\alpha] = \frac{(2 - \alpha)2 + 2}{12} = \frac{3 - \alpha}{6}. \tag{8.17}$$

Hence, from (8.17) the variance of a slope random variable Y is given by

$$Var[Y] = E[Y^2|\alpha] - E^2[Y|\alpha] = \frac{3 - (1 - \alpha)^2}{36}. \tag{8.18}$$

Note also that $\alpha = 1$ leads to the familiar mean and variance values $1/2$ and $1/12$ for a uniform$[0, 1]$ distribution.

Next, utilizing the mean of a slope variable Y (8.17) and the general expression for the mean of a TS variable (8.11), we derive the mean of a ST-SS variable X to be:

$$E[X|\theta, \alpha] = \frac{2(1 - \alpha)\theta + (\alpha + 2)}{6}, \tag{8.19}$$

and using the variance of a slope variable (8.18) and the general expression for the variance of a ST-SS variable (8.12), the variance of X becomes

$$Var(X|\theta, \alpha) = \tag{8.20}$$
$$\frac{3 - (1 - \alpha)^2}{36} - \frac{(2\alpha + 1)(1 - \alpha)}{18}\theta(1 - \theta).$$

Figures 8.4 and 8.5 plot the mean value $E[X|\theta, \alpha]$ (8.19) and the variance $Var(X|\theta, \alpha)$ (8.20) as a function of the threshold parameter θ for different values of α (the values of α in Figs. 8.4 and 8.5 coincide with those of the pdf's plotted in Fig. 8.1). From Fig. 8.4 it follows that the mean values of unimodal ST-SS distributions (Fig. 8.1B) are located consistently between the mean values of a triangular distribution (Fig. 8.1A) and a uniform one (Fig. 8.1C) with the triangular mean being the smaller (the larger) one for $0 \leq \theta \leq \frac{1}{2}$ ($\frac{1}{2} \leq \theta \leq 1$). Conversely, the mean values of U-shaped ST-SS distributions (Fig. 8.1D) also consistently fall between the mean values of a reverse triangular distribution (Fig 8.1E) and a uniform one (Fig. 8.1C) with the reverse triangular mean being the smaller (the larger) one for $0 \leq \theta \leq \frac{1}{2}$ ($\frac{1}{2} \leq \theta \leq 1$).

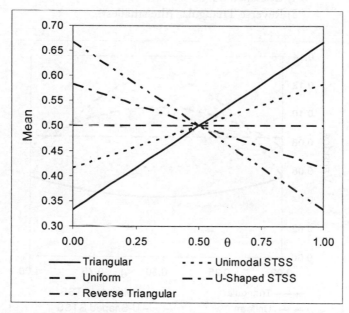

Fig. 8.4 Mean value (8.19) of ST-SS pdf's for different values of the slope parameter α.
A: $\alpha = 0$ (triangular), B. $\alpha = 0.5$ (unimodal ST-SS), C: $\alpha = 1$ (uniform),
D: $\alpha = 1.5$ (U-shaped ST-SS), E: $\alpha = 2$ (reverse triangular).

To summarize, for $0 \leq \theta \leq \frac{1}{2}$ the ordering of the mean values in descending manner is:

1) Reverse Triangular (the largest)
2) U-shaped ST-SS
3) Uniform
4) Unimodal ST-SS
5) Triangular (the smallest),

while for $\frac{1}{2} \le \theta \le 1$ we have conversely:

1) Triangular (the largest)
2) Unimodal ST-SS
3) Uniform
4) U-shaped ST-SS
5) Reverse Triangular (the smallest).

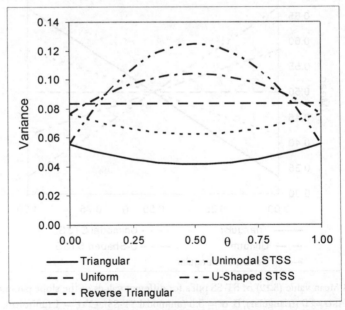

Fig. 8.5 Variance value of ST-SS distributions (8.20) for different values of the slope parameter α. A: $\alpha = 0$ (triangular), B: $\alpha = 0.5$ (unimodal ST-SS), C: $\alpha = 1$ (uniform), D: $\alpha = 1.5$ (U-shaped ST-SS), E: $\alpha = 2$ (reverse triangular).

Similar conclusions can be drawn for the variance from Fig. 8.5. Specifically, it follows from Fig. 8.5 that, for unimodal ST-SS distributions ($0 \leq \alpha \leq 1$), the variance falls between that of the uniform one and the triangular one (with the uniform distribution having the higher variance). Also note, that the minimal variance of unimodal ST-SS distributions is attained for $\theta = \frac{1}{2}$ (with the value of $1/24 \approx 0.042$ in the case of a symmetric triangular distribution). This cannot be said about the U-shaped ST-SS distributions ($1 \leq \alpha \leq 2$). In this case the maximum variance is attained at $\theta = \frac{1}{2}$ (with a value of $1/8 = 0.125$ for a symmetric reverse triangular distribution while dipping to the value $1/18 \approx 0.056$ at $\theta = 0$ or $\theta = 1$). In the vicinity of $\theta = \frac{1}{2}$ the ordering of variances is the same as the ordering of the mean values for $0 \leq \theta \leq \frac{1}{2}$ (Compare with p. 238). Thus the ordering of variances as a function of α is <u>not</u> preserved for the entire domain $\theta \in [0, 1]$. Also note, that a U-shaped ST-SS distribution may have a larger or smaller variance than that of the uniform distribution depending on the value of the reflection parameter θ (Compare with Fig. 8.5).

By means of a linear transformation $Z = (b - a)X + a$ we obtain from (8.15) the T-SS pdf:

$$
g\{z|m, \alpha, a, b\} = \begin{cases} \frac{\alpha}{b-a} + \frac{2(1-\alpha)}{b-a} \frac{z-a}{m-a}, & \text{for } a < z \leq m \\ \frac{\alpha}{b-a} + \frac{2(1-\alpha)}{b-a} \frac{b-z}{b-m}, & \text{for } m < z < b, \end{cases}
$$

and from (8.16) the T-SS cdf:

$$
G(z|m, \alpha, a, b) =
$$
$$
\begin{cases} \alpha \frac{z-a}{b-a} + (1-\alpha) \frac{b-a}{m-a} \left(\frac{z-a}{m-a} \right)^2, & \text{for } a < z \leq m \\ 1 - \alpha \frac{b-z}{b-a} - (1-\alpha) \frac{b-a}{b-m} \left(\frac{b-z}{b-m} \right)^2, & \text{for } m < z < b. \end{cases}
$$

where $0 \leq \alpha \leq 2$, $a \leq m \leq b$ and $m = (b - a)\theta + a$. Similarly we derive from (8.19) the mean value

$$
E[Z|m, \alpha, a, b] = \frac{\alpha + 2}{6} a + \frac{(1 - \alpha)}{3} m + \frac{\alpha + 2}{6} b \tag{8.21}
$$

and from (8.20) the variance of T-SS variable Z

$$Var(Z|m, \alpha, a, b) = \frac{3 - (1 - \alpha)^2}{36}(b - a)^2 - \tag{8.22}$$
$$\frac{(2\alpha + 1)(1 - \alpha)}{18}(m - a)(b - m).$$

Note that similar to the mean value formula of a four-parameter TSP distribution (Eq. (6.7), Sec. 6.1, Chapter 6) the weight of the most likely value m in (8.21) becomes negative when the pdf takes on a U-shaped form ($1 < \alpha \le 2$ for T-SS distributions and $0 < n < 1$ for TSP distributions). From (8.21) and (8.22) we obtain the mean and the variance of a uniform distribution (triangular distribution) with support $[a, b]$ by setting $\alpha = 1$ ($\alpha = 0$). For $\alpha = 2$, the mean and the variance of Z are reduced to

$$E[Z|m, 2, a, b] = \frac{2}{3}a - \frac{1}{3}m + \frac{2}{3}b$$

$$Var(Z|m, 2, a, b) = \frac{1}{18}(b - a)^2 + \frac{5}{18}(m - a)(b - m),$$

respectively. Observe that the "weight" for the mixing term $(m - a)(b - m)$ in the expression for $Var(Z|m, 2, a, b)$ is substantially higher than for the quadratic term $(b - a)^2$.

8.2.1 Moment estimation for ST-SS distributions

We now turn to a discussion of estimation procedures for the parameters of an ST-SS distribution. The next Sec. 8.2.2 describes the corresponding ML estimation which, as usual in this book, contains subtle and non-trivial arguments. Let for a sample $\underline{X} = (X_1, \ldots, X_n)$ of size n (not to be confused with the parameter n of the ogive pdf (8.11) or the same parameter of the power pdf (8.3) and generalized power pdf (8.5)) from a ST-SS pdf (8.15) the sample mean and the sample variance be given by the standard expressions

$$\overline{x} = \frac{1}{n}\sum_{i=1}^{n}X_i \text{ and } \widehat{\sigma}^2 = \frac{1}{n-1}\sum_{i=1}^{n}(X_i - \overline{x})^2. \tag{8.23}$$

Equating the population mean and variance (8.19) and (8.20), to their sample equivalents (8.23) we arrive at the following set of two equations needed to be solved in terms of θ $(0 \leq \theta \leq 1)$ and α $(0 \leq \alpha \leq 2)$:

$$\begin{cases} \frac{2(1-\alpha)\theta+(\alpha+2)}{6} = \overline{x} \\ \frac{3-(1-\alpha)^2}{36} - \frac{(2\alpha+1)(1-\alpha)}{18}\theta(1-\theta) = \widehat{\sigma}^2. \end{cases} \qquad (8.24)$$

From (8.24) and setting

$$\beta = 1 - \alpha \qquad (8.25)$$

we obtain the following quadratic equation which ought to be solved in terms of β:

$$\beta^2 - \{2 - 6(2\overline{x}-1)^2 - 24\widehat{\sigma}^2\}\beta - 9(2\overline{x}-1)^2 = 0 \qquad (8.26)$$

(where $-1 \leq \beta \leq 1$). Now θ follows from the first equation of (8.24) to be:

$$\theta = \frac{6\overline{x} - \alpha - 2}{2(1-\alpha)} \qquad (8.27)$$

(after substituting the estimated value of $\alpha = 1 - \beta$). Alternatively, a moment estimator of θ can be deduced by solving the quadratic equation in θ using the second part of (8.24). Note that from Figs. 8.4 and 8.5 it follows that necessary conditions for the existence of a solution to the moment equations (8.24) are:

$$0 \leq \overline{x} \leq 1 \text{ and } \frac{1}{24} \leq \widehat{\sigma}^2 \leq \frac{1}{8}. \qquad (8.28)$$

We shall illustrate the method of moments procedure for ST-SS distributions by providing four examples. Specifically, the uniform case ($\overline{x} = \frac{1}{2}$, $\widehat{\sigma}^2 = 1/12$, Fig. 8.1C), a symmetric triangular case ($\overline{x} = \frac{1}{2}$ and $\widehat{\sigma}^2 = 1/24$), a symmetric reverse triangular case ($\overline{x} = \frac{1}{2}$ and $\widehat{\sigma}^2 = 1/8$) and as the fourth example the illustrative sample

$$(X_{(1)}, \ldots, X_{(8)}) = (0.10, 0.25, 0.30, 0.40, 0.45, 0.60, 0.75, 0.80) \qquad (8.29)$$

(which was already used as example (1.38) in Chapter 1 and Chapter 3, Sec. 3.3.1).

Example 1: Setting $\overline{x} = \frac{1}{2}$ and $s^2 = 1/12$ (the mean and the variance of a uniform distribution), the quadratic equation (8.26) reduces to

$$\beta^2 = 0,$$

and from (8.25) it immediately follows that $\alpha = 1$ coinciding with the uniform member of the two-sided slope family (Eq. (8.15) and Fig. 8.2C).

Example 2: Setting $\overline{x} = \frac{1}{2}$ and $s^2 = 1/24$ (the mean and the variance of a symmetric triangular distribution), the quadratic equation (8.26) reduces now to

$$\beta^2 - \beta = 0 \Leftrightarrow \beta = 0 \text{ and } \beta = 1,$$

and from (8.25) it immediately follows that $\alpha = 1$ or $\alpha = 0$. The solution $\alpha = 1$ should be ruled out in view of the value $s^2 = 1/24$; substituting $\overline{x} = \frac{1}{2}$, $\alpha = 0$ into (8.27) we arrive at $\theta = \frac{1}{2}$.

Example 3: Setting $\overline{x} = \frac{1}{2}$ and $s^2 = 1/8$ (the mean and the variance of a symmetric reverse triangular distribution), the quadratic equation (8.26) reduces here to

$$\beta^2 + \beta = 0 \Leftrightarrow \beta = 0 \text{ and } \beta = -1,$$

and from (8.25) $\alpha = 1$ or $\alpha = 2$. The solution $\alpha = 1$ should be ruled out in view of the value $s^2 = 1/8$ and substituting $\overline{x} = \frac{1}{2}$, $\alpha = 2$ into (8.47) we arrive at $\theta = \frac{1}{2}$.

Example 4: From the sample values given in (8.29) and (8.23) it immediately follows that in this case

$$\overline{x} = 4.563e - 1 \text{ and } s^2 = 6.031e - 2. \tag{8.30}$$

Utilizing (8.26) and (8.30) we have the following quadratic equation in β:
$$\beta^2 - 0.507\beta - 0.069 = 0,$$
with the solutions

$$\beta_1 = -0.111 \text{ and } \beta_2 = 0.519.$$

Hence from (8.25) and (8.27) we obtain

$$\alpha_1 = 1.111, \theta_1 = 1.677 \text{ and } \alpha_2 = 0.481, \theta_2 = 0.247. \tag{8.31}$$

The first solution in (8.31) should be ruled out in view of the restriction $0 \le \theta \le 1$.

8.2.2 Maximum likelihood estimation of ST-SS parameters

While the method of moments for ST-SS distribution is quite elegant and straightforward, the ML estimation procedure may pose some technical difficulties. Let for a random i.i.d. sample of size n, $(X_1, \ldots, X_n)$, the corresponding order statistics be $X_{(1)} < X_{(2)} \ldots < X_{(n)}$. The likelihood for X with distribution (8.15) is by definition

$$L(\underline{X}; \theta, \alpha) = \tag{8.32}$$

$$\prod_{i=1}^{r} \left(\alpha + 2(1 - \alpha)\frac{X_{(i)}}{\theta} \right) \times \prod_{i=r+1}^{n} \left(\alpha + 2(1 - \alpha)\frac{1 - X_{(i)}}{1 - \theta} \right)$$

where $X_{(r)} \le \theta \le X_{(r+1)}$ with $X_{(0)} \equiv 0$, $X_{(s+1)} \equiv 1$. The likelihood (8.32) for the data (8.29) as a function of α and θ is depicted in Fig. 8.6A and its log-likelihood is displayed in Fig. 8.6B. Note that a global maximum of the likelihood can be clearly observed visually in Fig. 8.6A (the same maximum may be observed in Fig. 8.6B although it is somewhat less obvious). We shall compare (8.32) with the form of the likelihood of the STSP distribution given by Eq. (3.26) and (3.27) in Chapter 3.

Recall that in the case of a STSP distribution (3.11) in Chapter 3 maximization of the likelihood turns out to be a separable procedure, allowing first maximization with respect to θ (via $(n + 1)$ maximizations over the disjoint intervals $[X_{(r)}, X_{(r-1)}]$) and next maximizing with respect to the power parameter of the STSP variable (denoted in Chapter 3 by n). The form of the likelihood (8.32) does not allow, unfortunately, to apply a similar procedure and we are required to maximize $L(\underline{X}; \alpha, \theta)$ with respect to both parameters α and θ over the $n + 1$ sets

$$\theta \in [X_{(r)}, X_{(r+1)}], 0 \le \alpha \le 2, \text{ where } r = 0, \ldots, n. \tag{8.33}$$

It can however be shown that for every fixed $\theta \in [0, 1]$ the log-likelihood $Log\{L(\underline{X}; \theta, \alpha)\}$ is in fact a concave function over $\alpha \in [0, 2]$ (see Eq.

(8.37)). Figure 8.7 depicts profiles of the log-likelihood obtained using (8.32) for the data (8.29) as a function of α for fixed θ. (Please note, that the scales of the y-axes in Figs. 8.7A-D are not the same.)

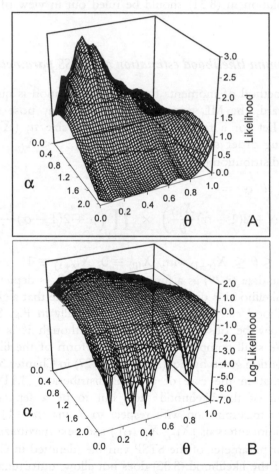

Fig. 8.6 A: The likelihood (8.32) as a function of α and θ for the data (8.29)
B: The log-likelihood of the likelihood (8.32) as a function of α and θ for the same data.

Hence, for a fixed θ the log-likelihood $L(\underline{X}\;;\theta,\alpha)$ is maximized over α at either $\alpha = 0$ (Fig. 8.7B) or $\alpha = 2$ or at the *unique* stationairy point (Figs.

242

8.7A, C and D) for which

$$\frac{\partial Log\{L(X\,;\alpha,\theta)\}}{\partial \alpha} = 0. \qquad (8.34)$$

In addition, since the sets (8.33) are bounded for $r = 0, \dots, n$ it follows that the unique global maxima $(\theta_{(r)}, \alpha_{(r)})$ over these sets do exist with the corresponding likelihood value $L(\underline{X}; \theta_{(r)}, \alpha_{(r)})$. Next, the ML estimators $\widehat{\theta}$ and $\widehat{\alpha}$ can be determined utilizing (as it has been done for the triangular and TSP distribution in Chapters 1, 3 and 4):

$$(\widehat{\theta}, \widehat{\alpha}) = \underset{r \in \{0, \dots, n\}}{arg\,max} \quad L(\underline{X}\,; \theta_{(r)}, \alpha_{(r)}). \qquad (8.35)$$

(See also Eqs. (1.30) and (1.31), Sec. 1.4 in Chapter 1.)

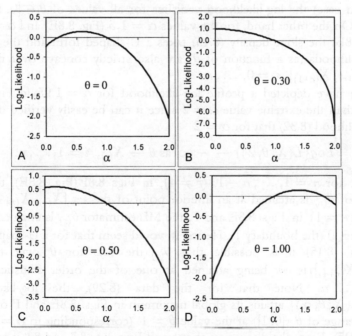

Fig. 8.7 Profiles of the log-likelihood based on (8.32) as a function of α for fixed θ for the data (8.29). A: $\theta = 0$; B: $\theta = X_{(3)} = 0.30$; C: $\theta = 0.50$; D: $\theta = 1.00$.

Figure 8.8 depicts profiles of the log-likelihood of (8.32) for the data (8.29) as a function of θ for a fixed α. Note (as before) that the scales of the y-axes in Figs. 8.8A-F differ. The order statistics $X_{(1)}, \ldots, X_{(n)}$ are indicated in Fig. 8.8 by vertical dotted lines (as is the case of Fig. 1.5 in Chapter 1). For the values $\alpha = 0$ (Fig. 8.8A), $\alpha = \frac{1}{3}$ (Fig. 8.8B) and $\alpha = \frac{2}{3}$ (Fig. 8.8C) the slope density (8.15) takes a unimodal form and the profile log-likelihoods are now reminiscent of the form of the function $H(\underline{X}|\theta)$ given by Eq. (1.24) displayed in Fig. 1.5 for the same data (8.29). (The function $H(\underline{X}|\theta)$ is also used to determine the ML estimator of θ in the ML procedure of the STSP distribution discussed in Chapter 3.) Observe that over the sets $\theta \in [X_{(r)}, X_{(r+1)}]$, $r = 0, \ldots, n$, the log-likelihood of (8.32) (analogously to the function $H(\underline{X}|\theta)$) is strictly convex in Figs. 8.8A, B and C and hence the ML estimator $\theta_{(r)}$ over these sets (for these three specific values of α) is attained either at $X_{(r)}$ or $X_{(r+1)}$. For $\alpha = 1$ (the uniform case) the log-likelihood vanishes for all values of $\theta \in [0, 1]$ (Fig. 8.8D). On the other hand, for the values $\alpha = 1.5$ (Fig. 8.8E) and $\alpha = 1.95$ (Fig. 8.8B) the slope density (8.15) takes a U-shaped form and the profile log-likelihoods (as a function of θ) are also strictly concave over the sets $\theta \in [X_{(r)}, X_{(r+1)}]$, $r = 0, \ldots, n$.

We have depicted a profile log-likelihood for $\alpha = 1.95$ in Fig. 8.8F rather than the extreme value $\alpha = 2$ since it can be easily verified utilizing the likelihood (8.32) that for $\alpha = 2$

$$Log\{L(\underline{X}; \theta, 2)\} \rightarrow -\infty \text{ as } \theta \rightarrow X_{(i)}, \ i = 1, \ldots, n.$$

Indeed, for $r = 1, \ldots, n - 1$, $(r \neq 4)$, in Figs 8.8F (Fig. 8.8E), the ML estimator $\theta_{(r)}$ is attained at an interior point of the set $[X_{(r)}, X_{(r+1)}]$. For $r = 0$ $(r = 1)$ in Figs. 8.8E and F, the ML estimator $\theta_{(r)}$ is attained at the boundary 0 (the boundary 1). Hence, it would seem that for a U-shaped ST-SS pdf (8.15) it is possible that $\theta_{(r)}$ (the ML for θ over the set $[X_{(r)}, X_{(r+1)}]$) is *not* being attained at one of the order statistics $X_{(i)}$, $i = 1, \ldots, n$. Note that for the data (8.29), the log-likelihood $Log\{L(\underline{X}; \theta, \alpha)\}$ attains its global maximum in Figs. 8.8E and F over the entire range of θ ([0, 1]) at the value $\theta = 1$ (corresponding to $\alpha = 1.5$ or $\alpha = 1.95$). From the discussion above and Figs. 8.6, 8.7 and 8.8 it can now be concluded that the ML estimators for $\theta \in [0, 1]$ and $\alpha \in [0, 2]$ for the data (8.29) are

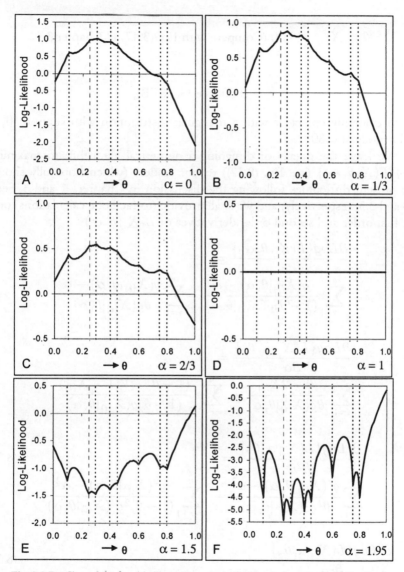

Fig. 8.8 Profiles of the log-likelihood based on (8.32) as a function of θ for fixed α for the data (8.29). A: $\alpha = 0$; B: $\alpha = \frac{1}{3}$; C: $\alpha = \frac{2}{3}$; D: $\alpha = 1$; E: $\alpha = 1.5$; F: $\alpha = 1.95$.

$$\widehat{\theta} = X_{(3)} = 0.30 \text{ (Compare with Eq. (3.47) in Chapter 3.)}$$

and

$$\widehat{\alpha} = 0.00 \text{ (see Fig. 8.7B)},$$

which evidently coincides with a triangular distribution with support $[0, 1]$ and a mode at 0.30.

We leave the details of designing a numerical algorithm to maximize $Log\{L(\underline{X}; \theta, \alpha)\}$ (see Eq. (8.32)) to our readers (which hopefully should not present difficulties following the discussion in Chapters 1 and 3, Secs. 1.4 and 3.3, respectively) and we shall provide (to facilitate calculations) only the first order and second order derivatives of $L(\underline{X}; \theta, \alpha)$:

$$\frac{\partial Log\{L(X; \theta, \alpha)\}}{\partial \theta} =$$

$$\sum_{i=r+1}^{n} \frac{g(X_{(i)}|\theta, \alpha) - \alpha}{(1 - \theta)g(X_{(i)}|\theta, \alpha)} - \sum_{i=1}^{r} \frac{g(X_{(i)}|\theta, \alpha) - \alpha}{\theta g(X_{(i)}|\theta, \alpha)},$$

$$\frac{\partial Log\{L(X; \alpha, \theta)\}}{\partial \alpha} =$$

$$\sum_{i=1}^{r} \frac{\theta - 2X_{(i)}}{\theta g(X_{(i)}|\theta, \alpha)} + \sum_{i=r+1}^{n} \frac{(1 - \theta) - 2(1 - X_{(i)})}{(1 - \theta)g(X_{(i)}|\theta, \alpha)},$$

$$\frac{\partial^2 Log\{L(X; \alpha, \theta)\}}{\partial^2 \theta} = \tag{8.36}$$

$$\sum_{i=1}^{r} \frac{g^2(X_{(i)}|\theta, \alpha) - \alpha^2}{\theta^2 g^2(X_{(i)}|\theta, \alpha)} + \sum_{i=r+1}^{n} \frac{g^2(X_{(i)}|\theta, \alpha) - \alpha^2}{(1 - \theta)^2 g^2(X_{(i)}|\theta, \alpha)},$$

$$\frac{\partial^2 Log\{L(X; \alpha, \theta)\}}{\partial^2 \alpha} = \tag{8.37}$$

$$-\sum_{i=1}^{r} \frac{\{\theta - 2X_{(i)}\}^2}{\{\theta g(X_{(i)}|\theta, \alpha)\}^2} - \sum_{i=r+1}^{n} \frac{\{(1 - \theta) - 2(1 - X_{(i)})\}^2}{\{(1 - \theta)g(X_{(i)}|\theta, \alpha)\}^2} < 0,$$

$$\frac{\partial Log\{L(\underline{X}\,;\alpha,\theta)\}}{\partial\theta\partial\alpha}=$$

$$\sum_{i=1}^{r}\frac{2X_{(i)}}{\{\theta g(X_{(i)}|\theta,\alpha)\}^2}+\sum_{i=r+1}^{n}\frac{2(1-X_{(i)})}{\{(1-\theta)g(X_{(i)}|\theta,\alpha)\}^2}>0.$$

The readers are of course encouraged to verify the first and the second order partial derivatives of the log-likelihood above. Note that for $0\le\alpha<1$ (corresponding to a uni-modal ST-SS pdf (8.15)) it follows from (8.36) and the fact that in this case the pdf $g(x|\theta,\alpha)\ge\alpha$ (namely α is the pdf value at the antimode θ) that

$$\frac{\partial^2 Log\{L(X\,;\alpha,\theta)\}}{\partial^2\theta}>0$$

over the sets $\theta\in[X_{(r)},X_{(r+1)}]$, $r=0,\ldots,n$. Hence, the log-likelihood $Log\{L(X\,;\alpha,\theta)\}$ is indeed strictly convex as a function of θ over these sets and the maximum is attained at one of the boundary points $X_{(r)}$ or $X_{(r+1)}$, $r=0,\ldots,n$ (for a fixed value of α). Conversely, for $1<\alpha\le2$ (corresponding to a U-shaped ST-SS pdf (8.15)) it follows from (8.36) and the fact that in this case the pdf $g(x|\theta,\alpha)\le\alpha$ (namely now α is the pdf value at the mode θ)

$$\frac{\partial^2 Log\{L(X\,;\alpha,\theta)\}}{\partial^2\theta}<0.$$

Hence now the log-likelihood $Log\{L(X\,;\alpha,\theta)\}$ is strictly concave as a function of θ over these sets and the maximum is attained at one of the boundary points $X_{(r)}$ or $X_{(r+1)}$, $r=0,\ldots,n$ (for a fixed value of α) or at a unique interior stationairy point of $Log\{L(X\,;\alpha,\theta)\}$ (in the interval $[X_{(r)},X_{(r+1)}]$, $r=0,\ldots,n$).

It remains so far an open research question whether the features of the ML procedure for ST-SS distributions described above (which are similar to those of the ML procedure for STSP distributions discussed in Chapter 3) can be extended to the general form of Standard Two-Sided families (Eq. (8.1)) with a monotonic generating density $p(\,\cdot\,|\Psi)$. A resolution of this problem may shed additional light on the ML procedures in non-standard cases.

Epilogue

To our readers: If you have studied this book carefully — and we are sure you did — you could not help but to learn a lot about an important class of statistical distributions, their structure, properties and methods of estimating parameters (some of them quite ingenuous) and applications. We sincerely hope that by now you are not "afraid" of statistical distributions and are comfortable with basic estimation methods. In addition, if you would encounter a new distribution (preferably a continuous one) we trust you will be happy to deal with it in confidence utilizing whenever possible modern computer and graphical methodology. We also hope that you will be able on your own to discover and investigate novel distributions that would be superior from statistical or engineering and scientific aspects (for the problem at hand) then the ones known so far.

You have also learned a bit about the history of some important "modeling" statistical distributions and by now appreciate the efforts of numerous scientists and statisticians who built this imposing edifice. Finally we trust that you may have improved your skills in mathematics and perhaps learned about applications of statistical methodology to some areas with which you are not quite familiar.

Our experience shows that many of our readers after diligently studying a book on specific distributions are left with a lingering question: *"What is the purpose of trying so hard to find a distribution which will fit accurately the data available to us?"* Suppose we are dealing with financial data (as we have done in this book on several occasions) regarding inherently uncertain phenomena. Conclusions about the behavior of these phenomena have therefore to be made having this uncertainty and variability in mind. Rather than drawing conclusions based on the variability in the empirical data alone, a close fit of a theoretical distribution would allow us further insight in the statistical properties of the phenomenon at hand. For example, could there be a practical justification for the discontinuity that was observed in the fit of the UTSP distribution to USA Certificate Deposit data from 1966-2002 discussed in Chapter 6, Sec. 6.5? It thus becomes clear that determining and fitting an appropriate distribution is not just a pleasant exercise in computational mathematics but (when properly designed and executed) can often shed a strong light on the phenomenon at hand and more importantly facilitate the discovery of the law governing this phenomenon.

You have no doubt observed that the text you have been studying does not contain exercises (partially substituted by comments and suggestions in the course of the exposition to carry out additional calculations and/or verify statements presented in the text). This is due to the novelty of the material which is scattered in numerous journals (financial, engineering and as well statistical) in the last 4-5 years starting from 1999 (a substantial part of it is from publications co-authored by the co-authors of this monograph). By compiling this book we intend to inaugurate a new — albeit modest — area in distribution theory containing continuous distribution on a bounded domain which possess meaningful parameters and attractive inferential properties (easily implemented using modern computational tools) that may have useful applications. We welcome your feedback and suggestions. This will assist us with the composition of a second edition which will hopefully contain a selection of appropriate exercises and some additional topics. We look forward to hearing from you and will be pleased to find out how useful this text is to you.

Appendix A

Graphical Overview of Continuous Univariate Families of Distributions possessing a Bounded Domain

Figure A.1 depicts graphically the various families of bounded distributions that are discussed in this book and is reminiscent of a similar graph developed by Leemis (1986) which also includes families of distributions with unbounded support. Amongst the distributions in Fig A.1 only the beta and uniform distributions are presented in Leemis (1986). The families indicated by shapes with rounded corners using dotted lines are discussed in Chapters 1-2 and of which its origins are dated pre-21 st century. Those indicated by sharp rectangles and solid lines are describes in Chapters 3-8 (appearing mostly in post 20-th century archival literature). The numbers in the corners of each rectangular (rounded or sharp) shape represent the number of parameters associated with family of distributions specified in it. Two of these parameters are used to identify the support of these families. Hence, if one were to consider the standardized versions with support $[0, 1]$ of these families of distributions one should subtract two (the lower and upper bound parameters) from the number of parameters indicated in each rectangle.

By starting at the rectangle of a particular family and following the arrows one can identify common members of parent families of distributions. Hence, both the beta and TSP families may be considered to be parent families of the power and reflected power distribution. The power and reflected-power families of distributions are in turn the parent families of the uniform distribution (in addition to the one-sided slope, distribution and the trapezoidal distribution). The dotted arrows indicate relationships that were earlier established (pre-21st century) and the solid arrows indicate relationships that have been identified more recently. One vividly observes from Fig. A.1 that the newly constructed univariate distributions with a bounded domain are more connected to the triangular (and uniform) distribution than the beta distribution and hence the title of this monograph. Table A.1 provides a roadmap for this book to most of the distributions presented in Fig. A.1.

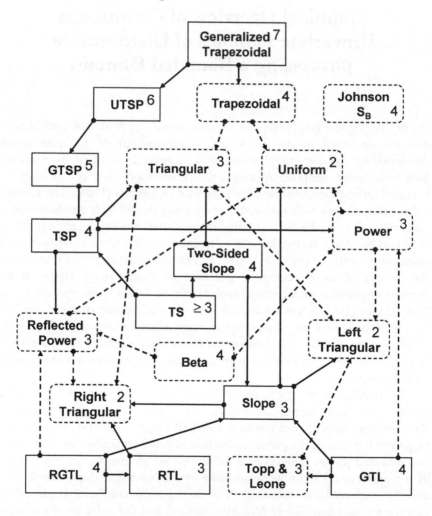

Fig. A.1 Overview of families of continuous univariate distributions on a bounded domain and their connections.

Table A.1 A roadmap for most of the families of distributions depicted in Fig. A.1.

	Acronym	Family of Distributions	# Parameters	Chapters	PDF Formula	Page
1	GT	Generalized Trapezoidal	7	5	(5.1)	150
2	LCUV	Linear Combination of Uniform Variables	≥ 1	2	(2.37)	50
3	STS	Standard Two-Sided	≥ 1	8	(8.1)	226
4	UTSP	Uneven Two-Sided Power	6	6	(6.1)	167
5	USTSP	Uneven Standard Two-Sided Power	4	6	(6.2)	168
6	GSTSP	Generalized Standard Two-Sided Power	5	6	(6.5)	169
7	T	Trapezoidal	4	2	(2.22)	45
8	STSP	Standard Two-Sided Power	2	3	(3.11)	71
9	TSP	Two-Sided Power	4	4	(4.2)	98
10		Standard Beta	2	1, 3	(1.5), (3.24)	4, 78
11		Beta	4	4	(4.1)	98
12	SRGTL	Standard Reflected Generalized Topp & Leone	2	7	(7.9)	201
13	RGTL	Reflected Generalized Topp & Leone	4	7	(7.4)	198
14	STL	Standard Topp & Leone	1	2	(2.5)	36
15	TL	Topp & Leone	2	2	(2.6)	37
16	ST-SS	Standard Two-Sided Slope	2	8	(8.15)	232
17	SS	Standard Slope	7,8	1	(7.1), (8.4)	196, 226
18	S	Slope	3	5	(5.6)	153
19	SP	Standard Power	1	8	(8.3)	226
20	P	Power	3	5	(5.5)	153
21	RP	Reflected Power	3	5, 7	(5.7), (7.7)	154, 199
22		Standard Triangular	1	1	(1.4)	4
23	Triang	Triangular	3	1	(1.6)	5
24	SRT	Standard Right Triangular	8	0	(8.2)	226
25	S_B	Johnson System S_B	4	Appendix B	(B.5), (B.6)	257

The Johnson S_B transformation yields a continuous unimodal distribution (or bimodal with an antimode between them) on $[0, 1]$. It is based on a logarithmic transformation of a Gaussian distribution and as such is not formally connected with the various families of distributions depicted in Fig. A.1 (although some graphs show a very close similarity). It may be of interest to compare analytically (or numerically) the interrelation – if any – between the Johnson's S_B distribution (see, e.g. Elderton and Johnson (1969)) and those depicted in Fig. A.1. Such a comparison to the best of our knowledge has not been carried out (partially due to historical reasons). Since the Johnson S_B is not connected to the other distributions in

Fig. A.1 a brief overview of this family is presented in Appendix B rather than the main part of the book. Given the popularity of the Johnson S_B distribution and its prominence we strongly feel it has a place in this book on other univariate continuous distributions with bounded support other than Pearson's beta distribution.

We conclude by noting that, in the last decade, integration of graphically interactive and statistical procedures for input distribution modeling has become a topic of intensive research (see, e.g., DeBrota *et al.* (1989), AbouRizk *et al.* (1991), Flanigan (1993) and Wagner and Wilson (1996)). This involves, amongst others, the software system PRIME and univariate Bézier curves (or distributions) which are a variant of spline functions (see Wagner and Wilson (1996)). Bézier curves utilize a number of *control points* as its parameters (the default number in PRIME is 6) where each control point is defined by an x and y coordinate. Two of these control points define its lower and upper bound. The remaining control points determine *domains of attraction* for the Bézier curve, but do not have to be points *on* the Bézier curve itself and may not shed additional light on the uncertainty phenomenon one is trying to model. A user may edit the location of a control point and, add or delete them which requires visual user interaction with the software PRIME. Alternatively, for a given number of control points, the software PRIME can numerically determine their location by minimizing its "distance" to available empirical data. An overview of a variety of distance measures is presented in Wagner and Wilson (1996). The system of Bézier distributions allows for great flexibility in input distribution modeling for stochastic simulations. However, as Wagner and Wilson (1996) mention, variate generation from a Bézier distribution is not computationally efficient (thus far) since its inverse cumulative distribution function is not available in a closed form (similar to the case of the beta, but unlike the TSP distributions).

Appendix B

The Johnson S_B Distribution

The reader is by now well familiar with the triangular distribution and some of its generalizations as possible alternatives for the beta distribution. N.L. Johnson (1917 -) who is one of the leaders of the 20-th century Statistics (specializing in Statistical Distribution Theory and Applications) and who studied in the Imperial College in London at the tail end of Karl Pearson's (the father of the beta distribution among other important distributions) life in the middle thirties of the 20-th century has developed in 1949 a system of transformations of the normal distribution which received substantial popularity in the second half of the 20-th century. The Johnson system contains the S_B class of distributions, which have a bounded support. Even though, the Johnson S_B distribution does not directly link to the other distributions presented in this book (see, Appendix A), it deserves consideration as a possible alternative to the beta distribution being a versatile continuous distribution on a bounded domain. It has been extensively investigated in the statistical literature for some 40 years. We shall provide here a brief overview.

B.1 Motivation and Representation

Suppose that berries of fruit have radii r (measured in some decidable fashion) normally distributed with parameters μ and σ^2. Then the surface area $S = \pi r^2$ will no longer be normal and indeed will be skewed distributionally. This idea was developed a little later than the introduction of the K. Pearson system of continuous distribution by J.C. Kapteyn (1903) in his treatment of skew frequency curves in biology. The development of the subject was perhaps retarded because of over anxiety and unsuccessful efforts to trace the transformation's relation to natural phenomena and mathematical difficulties. Moreover, Kapteyn's memoir proved to be intractable mathematically in most cases.

255

Actually, the idea of transforming from one random variate to another with a more convenient density has been developed in the first half of the 20-th century. If the distribution of the random variable Y is such that a simple transformation of Y has a well known distribution, it becomes possible to use the research on the latter — including published tables — in studying the former distribution. A pioneering step in this direction was carried out by N.L. Johnson in the above mentioned 1949 publication in *Biometrika*. Let X be a standard normal distribution with pdf

$$f(x) = \frac{1}{\sqrt{2\pi}} e^{-\frac{1}{2}x^2} \tag{B.1}$$

and consider the transformation

$$Y = g^{-1}\left(\frac{X - \gamma}{\delta}\right) \quad \text{or} \quad X = \gamma + \delta g(Y) \tag{B.2}$$

for a suitable function $g(\cdot)$ and parameters $\gamma \in \mathbb{R}$ and $\delta > 0$. The transformation of Y given by $\gamma + \delta g(Y)$ has the classical standard Gaussian distribution (B.1). We have for the density function of Y

$$h(y|\gamma, \delta) = \delta f(\gamma + \delta g(y))\left|\frac{dg(y)}{dy}\right| = \frac{\delta}{\sqrt{2\pi}} e^{-\frac{1}{2}\{\gamma + \delta g(y)\}^2}\left|\frac{dg(y)}{dy}\right|.$$

Johnson proposes three types of systems:
1) The S_L or lognormal system:

$$g(y) = Log(y);$$

2) The S_B (bounded support system):

$$g(y) = Log\{y/(1 - y)\} \tag{B.3}$$

3) The S_U (unbounded support):

$$g(y) = sinh^{-1}(y) = Log(y + \sqrt{y^2 + 1}).$$

Other systems, of course, are possible. However, the variety of shapes given by these three is quite as large as that of the whole Pearson system (see, e.g., Stuart and Ord (1994)): The S_B and S_U systems occupy non-overlapping regions covering the whole (β_1, β_2) plane; the S_L system defines the curve that separates them — the quantities β_1, β_2 are the familiar skewness and kurtosis respectively defined in Eq. (1.18) in Chapter 1). The system S_B which is of interest in our case results in the pdf:

$$h(y|\gamma, \delta) = \frac{\delta}{\sqrt{2\pi}} \frac{1}{y(1-y)} exp\left[-\frac{1}{2}\left\{ \gamma + \delta Log\left(\frac{y}{1-y}\right) \right\}^2 \right], \text{ (B.4)}$$

where $0 \leq y \leq 1$, $\delta > 0$. Similarly to the Gaussian distribution, the cdf of the Johnson S_B variable Y given by (B.0) is not available in a closed form. Using a simple linear scale transformation $Z = (b-a)Y + b$ we obtain the Johnson S_B pdf with support $[a, b]$ given by

$$h(z|a, b, \gamma, \delta) = \text{ (B.5)}$$
$$\frac{\delta}{\sqrt{2\pi}} \frac{(b-a)}{(z-a)(b-z)} exp\left[-\frac{1}{2}\left\{ \gamma + \delta Log\left(\frac{z-a}{b-z}\right) \right\}^2 \right],$$

where $a \leq z \leq b$, $\gamma \in \mathbb{R}$ and $\delta > 0$. The Johnson S_B system has found application in fields like meteorology (see, Johnson, 1949), medicine (see, e.g., Johnson (1949), Bukac (1972)), biology (see, e.g., Draper (1952), Slifker and Shapiro (1980)), forestry (see Kudus *et al.* (1999)) and other sciences (see Mage (1980)) and due to its flexibility serves as an important alternative to the beta distribution in modeling input distributions for stochastic simulations (see, e.g., DeBrota *et al.* (1989)).

B.2 Some Properties of the Johnson S_B Family

Figure B.1 depicts some examples of the Johnson S_B pdf's and demonstrates the variety of forms that it may have. For $\gamma = 0$ the density function is symmetric on $[0, 1]$ (Figs. B.1A and B). This follows immediately from rewriting the pdf (B.4) in a form that is reminiscent of the beta pdf (1.18) in Chapter 1, yielding

$$h(y|\gamma, \delta) = \frac{\delta}{\sqrt{2\pi e^{\gamma^2}}} y^{-\zeta(y|\delta,\gamma)-1}(1-y)^{\zeta(y|\delta,\gamma)-1}, \text{ (B.6)}$$

where $\delta > 0$ and

$$\zeta(y|\gamma, \delta) = \frac{\delta^2}{2}g(y) + \gamma\delta, \text{ (B.7)}$$

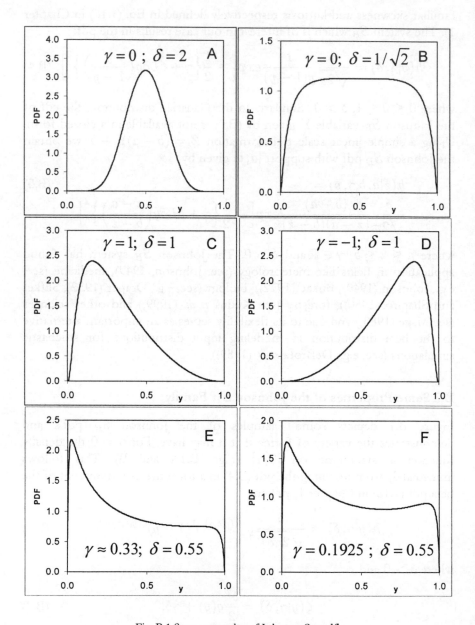

Fig. B.1 Some examples of Johnson S_B pdf's.

where $g(y)$ is the original Johnson S_B transformation function given by (B.3). In fact, denoting the pdf of $Z = 1 - Y$ (i.e. the reflection of Y) by $f(z|\gamma, \delta)$ we obtain from (B.6) and (B.7) that

$$f(z|\gamma, \delta) = h(z| - \gamma, \delta). \tag{B.8}$$

Hence from (B.8) it follows that Fig. B.1D is the *reflection* of Fig. B.1.C. and similarly to the beta (Eq. (1.5) in Chapter 1) and STSP (Eq. (3.11) in Chapter 3) distributions, the reflected Johnson S_B distribution is also a member of the Johnson S_B family.

The density function (B.4) may be bimodal (see, Fig. B.1F). Note that, while the Johnson S_B pdf takes on somewhat of a U-Shaped form, the modes are not attained at the support boundaries (unlike the case of beta distribution and STSP distributions) since the density value of a Johnson S_B distribution at the boundaries 0 and 1 remains 0, regardless of the values of the parameters δ and γ in (B.6). This assertion follows from (B.6) and (B.7) and the fact that $\zeta(y|\gamma, \delta) \to \infty$ ($\zeta(y|\gamma, \delta) \to -\infty$) when $y \uparrow 1$ ($y \downarrow 0$). We shall conduct a detailed mode analysis in a subsection below.

B.2.1 Median value

Since the median of a standard normal random variable is at zero we immediately obtain from (B.2) the median $y_{0.5}$ of Y for the Johnson system to be

$$y_{0.5} = g^{-1}\left(\frac{-\gamma}{\delta}\right), \tag{B.9}$$

where $g^{-1}(\,\cdot\,)$ is the inverse function of $g(\,\cdot\,)$. In case of the Johnson S_B system we have from (B.3) that

$$g^{-1}(y) = \frac{1}{1 + e^{-y}}, \tag{B.10}$$

and using (B.10) the median becomes

$$y_{0.5} = \frac{1}{1 + e^{\frac{\gamma}{\delta}}} \tag{B.11}$$

for the Johnson S_B family. The RHS of (B.10) is the well-known sigmoid or standard logistic function (see, e.g., Von Seggern (1993)) and should *not* be confused here with the cdf of a standard logistic distribution (see, e.g., Johnson *et al.* (1995)) which has the same functional form (and an unbounded support).

In Figs. B.1C and B.1D we have the medians to be respectively:

$$y_{0.5} = \frac{1}{1+e} \text{ and } y_{0.5} = \frac{1}{1+e^{-1}} = 1 - \frac{1}{1+e},$$

which also follows from the fact (as mentioned above) that the pdf in Fig. B.1D is the reflection of the pdf in Fig. B.1C. Note that from (B.11) it follows that the ratio

$$\kappa = \frac{\gamma}{\delta} \tag{B.12}$$

may be interpreted as a *location ratio*.

B.2.2 Mode analysis

It is of interest to describe the possible location(s) of the mode(s) of the Johnson S_B family. Differentiating (B.6) and equating to zero we have

$$2y - 1 = \delta\left\{\gamma + \delta Log\left(\frac{y}{1-y}\right)\right\} \tag{B.13}$$

or using the transformation

$$z = 2y - 1 \tag{B.14}$$

and the definition of κ given by (B.12), we arrive at

$$\frac{z}{\delta^2} - \kappa = Log\left(\frac{1+z}{1-z}\right), \tag{B.15}$$

where $z \in [-1, 1]$. The LHS of Eq. (B.15) is a strictly increasing straight line with the slope solely determined by the parameter δ (see, Fig. B.2). The derivative of the RHS of (B.15) equals

$$\frac{2}{1-z^2} \tag{B.16}$$

which attains its minimal value of 2 at $z = 0$. Setting (B.16) to be $1/\delta^2$ we obtain the following two equations

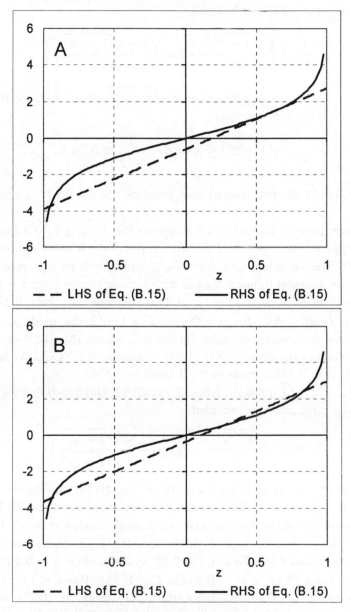

Fig. B.2 Graphical depiction of Eq. (B.15).
A: $\delta = 0.55$, $\gamma = 0.33$ (see Fig. B.1E); B: A: $\delta = 0.55$, $\gamma = 0.1925$. (see Fig. B.1F)

$$\frac{1}{\delta^2}z \pm \left\{ \frac{\sqrt{1-2\delta^2}}{\delta^2} - Log\left(\frac{1+\sqrt{1-2\delta^2}}{1-\sqrt{1-2\delta^2}}\right) \right\} \qquad \text{(B.17)}$$

for the possible two *tangent* lines at $\pm\sqrt{1-2\delta^2}$ of the RHS of function (B.15) with slope $1/\delta^2$ provided

$$\sqrt{1-2\delta^2} \leq 1 \Leftrightarrow \delta \leq \frac{1}{\sqrt{2}} \approx 0.70. \qquad \text{(B.18)}.$$

For $\delta = 1/\sqrt{2}$ the two tangent lines coincide and Eq. (B.17) simplifies to $z/2$.

Hence, from (B.15) and (B.17) it follows that for $\delta \geq 1/\sqrt{2}$ Eq. (B.15) has a single solution and the Johnson S_B pdf (B.6) is unimodal (see Figs. B.1A-D). The boundary case $\delta = 1/\sqrt{2}$ and $\kappa = 0$ ($\gamma = 0$) results in a symmetric unimodal Johnson S_B pdf with a mode at $y = \frac{1}{2}$ (or $z = 0$ using the transformation $y = 2z - 1$) and is depicted in Fig. B.1B. For $\delta > 1/\sqrt{2}$ and $\gamma \neq 0$ (for $\gamma = 0$ and $\delta > 1/\sqrt{2}$ the mode is also at $y = \frac{1}{2}$), one can numerically solve for the mode from Eq. (B.15) using, e.g., the root-finding algorithm GOALSEEK in Microsoft EXCEL. We have for Fig. B.1C (Fig. B.1D) a mode at ≈ 0.156 (≈ 0.854).

For $\delta < 1/\sqrt{2}$ it follows from (B.17), (B.12) and $\delta > 0$ that Eq. (B.15) has two or three solutions provided

$$|\gamma| \leq \frac{\sqrt{1-2\delta^2}}{\delta} - \delta Log\left(\frac{1+\sqrt{1-2\delta^2}}{1-\sqrt{1-2\delta^2}}\right). \qquad \text{(B.19)}$$

When (B.19) attains equality, the LHS of Eq. (B.15) coincides with the tangent line (B.17) of its RHS resulting in two solutions of Eq. (B.15) and the Johnson S_B pdf is *still* unimodal with a single mode less than $\frac{1}{2}$ (greater than $\frac{1}{2}$) when $\gamma > 0$ ($\gamma < 0$). Figures B.2A and B.1E depict this case for the specific values $\delta = 0.55$ and $\gamma \approx 0.33$ with a mode at $y \approx 0.023$ in Fig. B.1E (or from (B.14) $z \approx -0.954$ in Fig. B.2A). For $\delta < 1/\sqrt{2}$ and a strict inequality of (B.19) it follows from (B.17) that Eq. (B.15) has three solutions and the Johnson S_B pdf is bimodal. Figures B.2B and B.1F depict this case for $\delta = 0.55$ and $\gamma \approx 0.192$ with modes at $y \approx 0.031$ and

$y \approx 0.198$ in Fig. B.1F (or from (B.14) $z \approx -0.939$ and $z \approx 0.835$, respectively in Fig. B.2B).

B.2.3 Moments and parameter estimation

The k-th non-central moment associated with the pdf (B.4) or (B.6) is given by

$$\mu_k' = E[Y^k] = \frac{1}{\sqrt{2\pi}} \int_{-\infty}^{\infty} e^{-\frac{u^2}{2}} \left\{ 1 + e^{\left(\frac{\gamma-u}{\delta}\right)} \right\}^k du, \quad k = 1, 2, \ldots \text{(B.20)}$$

(see, Johnson, 1949) which is a transcendental quantity that is somewhat difficult to evaluate. Johnson (1949) produced a formula for the mean $\mu_1' = E[Y]$ as the ratio of an infinite series involving the Jacobi theta function (not involving integrals). Johnson and Kitchen (1971a,b) also derived the following recursive relations for the non-central moments

$$\mu_k'(\gamma+\delta^{-1}, \delta) = e^{-\left(\frac{2}{\delta^2}+\frac{7}{\delta}\right)} \times \left\{ \mu_{k-1}'(\gamma, \delta) - \mu_k'(\gamma, \delta) \right\},$$

which were utilized to construct tables to facilitate fitting of the Johnson S_B distribution via the method of moments. Similarly to the Reflected Generalized Topp Leone distribution (discussed in Chapter 7) method of moments estimation is slightly cumbersome due to the structural form of (B.20), but has successfully been investigated Hill *et al.* (1976) and by Bacon-Shone (1985) for a five-parameter generalization of the Johnson S_B pdf (B.5).

Maximum likelihood estimation have been investigated rather recently by Kottegoda (1987) who noticed feasibility of the ML procedure for Johnson S_B distribution when kurtosis is low and difficulties in other cases. Difficulties associated with the ML and method of moment procedures have resulted in considerable efforts to propose alternative methods and to investigate their properties. Amongst them are least squares estimation by Swain *et al.* (1988), estimation of parameters based on percentile points (see, e.g., Johnson (1949), Bukac (1972), Mage (1980), Slifker and Shapiro (1980), Wheeler (1980), Bowman and Shenton (1988), Siekierski (1992), Zhou and McTague (1996)). Even more recently Kudus *et al.* (1999) discuss a non-linear regression approach similar to the one in Swain *et al.* (1988) (although Kudus *et al.* (1999) do not refer to Swain *et al.* (1988) — at least not directly) and concluded that the non-linear regression method compares favorably compared to other ones discussed in their paper (which included the ML method). Since the least squares estimation procedure of Swain *et al.* (1988) can be straightforwardly implemented utilizing the SOLVER add-in in Microsoft EXCEL we shall provide a brief overview.

For a random sample $\underline{Y} = (Y_1, \ldots, Y_n)$ of size n from a Johnson S_B distribution let the order statistics be $Y_{(1)} < Y_{(2)} \ldots < Y_{(s)}$. From (B.1) and (B.2) it follows that

$$Pr(Y \leq Y_{(i)}) = F(Y_{(i)}|\gamma, \delta) = \Phi\{\gamma + \delta g(Y_{(i)})\}, \qquad (B.21)$$

where $\Phi\{\cdot\}$ is the standard normal cdf (for which, for example, a standard function NORMDIST is available in Microsoft EXCEL) and $g(\cdot)$ for the Johnson S_B system is defined by (B.3). Swain *et al.* (1988) shows that

$$F(Y_{(i)}|\gamma, \delta) = \frac{i}{n+1} + \epsilon_i, 1 \leq i \leq n, \qquad (B.22)$$

where for the residual terms ϵ_i

$$E[\epsilon_i] = 0, i = 1, \ldots, n$$

and

$$Cov(\epsilon_i, \epsilon_k) = \frac{1}{n+2}\left\{\frac{i}{n+1}\left(1 - \frac{k}{n+1}\right)\right\} > 0,$$

for $i \leq k, k = 1, \ldots, n$. Next, using weighted non-linear regression to solve for the parameters γ and δ, one minimizes

$$\sum_{i=1}^{n} w_i\left\{F(Y_{(i)}|\gamma, \delta) - \frac{i}{n+1}\right\}^2 \qquad (B.23)$$

for some weights $w_i > 0$ using standard optimization tools, such as, e.g., the SOLVER add-in in Microsoft EXCEL under the restriction that $\delta > 0$. Setting $w_i = 1, i = 1, \ldots, n$ is referred to as the *ordinary least squares (OLS)* method, while setting

$$w_i = \frac{1}{Var(\epsilon_i)} = \frac{n+2}{\frac{i}{n+1}\left(1 - \frac{i}{n+1}\right)}$$

is referred to as the *Diagonal Weighted Least Squares* (DWLS) method (see, e.g. AbouRizk (1990). The latter method assigns a lower weight to the lower and upper order statistics since these have a larger variance than those observed in the middle ranges.

It is important to note that the above estimation procedure may be straightforwardly extended to include estimation of the lower and upper bounds parameters a and b of the Johnson S_B distribution (B.5), by adding the constraints

$$a \leq X_{(1)} \text{ and } b \geq X_{(s)}$$

to the optimization procedure minimizing (B.23). Moreover, the above estimation procedure may be extended to any arbitrary cdf by appropriate modification of (B.21). In fact, the OLS estimation procedure above (see also Eq. (4.20) in Chapter 4) was utilized to fit a four parameter beta distribution (see Eq. (4.1) in Chapter 4) to the data in Table 1.2 in Chapter 1 to avoid possible numerical difficulties with the four-parameter ML procedure for the beta distribution (see Carnahan (1989)).

B.2.4 Some thoughts on limiting distributions

To the best of our knowledge limiting distribution of the Johnson S_B distribution have not been formally investigated, perhaps due to the difficulties involved with calculating the moments. We would like to offer some thoughts here that are primarily based on visual observations. We have depicted in Fig. B.3 some additional examples of the Johnson S_B to help us collect these thoughts.

From Fig. B.3A, (B.8) and the expression for the median (B.11) it seems that when $\gamma \to \infty$ ($\gamma \to -\infty$), keeping δ fixed that the pdf (B.6) converges to a single point mass at 0 (at 1). Similarly, letting $\gamma \to \infty$, keeping $\kappa = \gamma/\delta$ defined by (B.12) constant it seems from (B.11) that the pdf (B.6) converges to a single point mass at

$$\{1 + e^{\kappa}\}^{-1}$$

or $\{1 + \sqrt{e}\}^{-1} \approx 0.378$ in case of Fig. B.3B.

Since the RHS of (B.19) becomes arbitrarily large when $\delta \downarrow 0$ it follows immediately that the Johnson S_B distribution becomes bimodal regardless of the value of γ (Recall that the density value of the pdf (B.6) always equals zero at the boundaries 0 and 1 for all parameter values γ and $\delta > 0$). From (B.2), X being standard normal distributed and the fact that the function $g(\,\cdot\,)$ given by (B.3) is strictly increasing, it follows immediately that

$$Pr(Y \le y) = \Phi\{\gamma + \delta g(y)\}, \tag{B.24}$$

where $\Phi(\,\cdot\,)$ is the standard normal cdf. Hence, utilizing (B.24) it would seem that as $\delta \downarrow 0$ (keeping γ fixed) the pdf (B.6) converges to a Bernoulli distribution with a probability mass $\Phi(\gamma)$ at 0 and $\{1 - \Phi(\gamma)\}$ at 1.

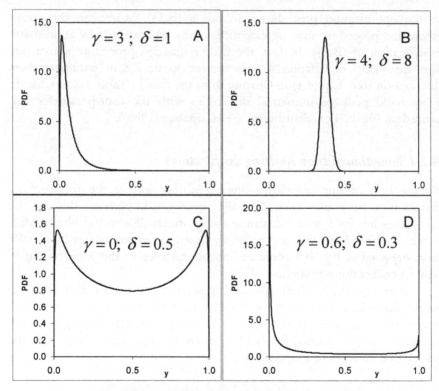

Fig. B.3 Some additional examples of Johnson S_B pdf's.

If the above assertions prove to be true, the Johnson S_B family enjoys the same limiting distributions as the beta and STSP families, which would add to its flexibility.

B.3 Concluding Remarks

It is an interesting that the pioneering Johnson's contributions to the development of an elegant and powerful system of transformations of the normal distribution (consisting of three families, which actually cover all of the twelve Pearson's curves — see, e.g., Patil *et al.* (1984)) were developed in the late forties of the 20-th century, before the introduction of (even the most primitive) computers into statistical practice. Computations related to the Johnson's system are quite ingenuous and involved and were originally carried out with the old-fashioned calculators which required long hours and even days of patient calculating of something that nowadays may take less than 1 second.

The family has no simple expression for the moments (which some 50 years ago may have been considered a serious drawback). However, the family has paved the way for introduction of computation intensive methodology and the use of non-standard functions (such as the hyperbolic and Jacobi functions) in statistical practice. In this sense it was certainly a substantial accomplishment which should in our opinion be definitely noted when dealing with continuous univariate distributions on a bounded domain. This system is still widely used for fitting curves in various fields of sciences, medicine and technology.

Bibliography

AbouRizk, S.M. (1990). *Input Modeling for Construction Simulation* (Phd. Thesis Purdue University). Ann Arbor, MI: UMI Dissertation Services.

AbouRizk, S.M., Halpin, D.W. and Wilson, J.R. (1991). Visual interactive fitting of beta distributions. *Journal of Constructions Engineering and Management,* 117 (4), 589-605.

AbouRizk, S.M. and Halpin, D.W. (1992). Statistical properties of construction duration Data. *Journal of Construction Engineering and Management,* 118 (3): 525-544.

Aigner, D.J., Amemiya, T., and Poirier, D. J. (1976). On the estimation of production frontiers: maximum likelihood estimation of the parameters of a discontinuous density function. *International Economic Review* 17 (2): 377-396.

Altiok, T. and Melamed, B. (2001). *Simulation Modeling and Analysis with Arena.* Cyber Research and Enterprise Technologies, pp. 46.

Alpert, M. and Raiffa, H. (1982). A progress report on the training of probability assessors, In *Judgment Under Uncertainty: Heuristics and Biases,* D. Kahneman, P. Slovic and A. Tversky (Eds.). New York, NY: Cambridge University Press, 294-305.

Ayyangar, A.S.K. (1941). The triangular distribution. *Mathematics Student,* 9: 85-87.

Bacon-Shone, J. (1985). Algorithm AS 210: Fitting five parameter Johnson SB curves by moments. *Applied Statistics,* 34: 95-101.

Balakrishnan, N. and Kocherlakota, S (1985). On the double Weibull distribution: order statistics and estimation. *Sankhyā,* 47 (B), 61-178.

Banks, J., Carson, J.S., Nelson, B.L., and Nicol, D.M. (2001). *Discrete-Event System Simulation* (3rd ed.). Upper Saddle River, NJ: Prentice-Hall.

Barabasi, A. (2003). *Linked: How everything is connected to everything else and what it means* (Reissue edition). New York, NY: Plume Publishers

Bardosi, G. and Fodor, J. (2004). *Evaluation of Uncertainty and Risks in Geology.* Berlin: Springer-Verlag

Barnard G. A. (1989). Sophisticated theory and practice in quality improvement. *Philosophical Transactions of The Royal Society* A, 327: 581-589.

Barrow, D.L. and Smith, P.W. (1979). Spline notation applied to a volume problem. *American Mathematical Monthly,* 86: 50-51.

Barsky, R.., Bound, J., Kerwin, K.C. and Lupton, J.P. (2002). Accounting for the black-white wealth Gap: A nonparametric approach. *Journal of the American Statistical Association*, 97 (459): 663-673.

Bartholomew, D.J., 1987. *Latent Variable Models and Factor Analysis*, Oxford University Press.

Basu, S. and Dasgupta, A. (1993). The mean, median and mode of unimodal distributions: a characterization. *Theory of Probability and its Applications*, 41 (2) 210:223.

Bates, G.E. (1955). Joint distributions of time intervals for the occurrence of successive accidents in a generalized Pólya scheme. *Annals of Mathematical Statistics*, 26, 705-720.

Bayes, T (1763). An essay towards solving a problem in the Doctrine of Chances. *Philosophical Transactions of the Royal Statistical Society in London*, 53: 370-418, Reproduced in *Biometrika*.

Brown, S.L. (1999). *An SAB Report: estimating Uncertainties in Radiogenic Cancer Risk*, Washington D.C.: Science and Advisory Board, United States Environmental Protection Agency, EPA-SAB-RAC-99-008.

Bowlus, A., Neumann, G. and Kiefer, A. (2001). Equilibrium search models and the transition from school to work. *International Economic Review*, 42 (2): 317-343.

Bowman, K.O. and Shenton, L.R. (1988). Solutions to Johnson's S_B and S_U. *Communication in Statistics*, 17 (2): 343-348.

Bukac, J. (1972). Fitting S_B curves using symmetrical percentile points, *Biometrika*, 59: 688-690.

Cardano, G. (1993). *Ars Magna or the Rules of Algebra* (English Translation). New York, NY: Dover Publications.

Carnahan, J.V. (1989). Maximum likelihood estimation for the 4-parameter beta distribution. *Communication in Statistics - Simulation and Computation*, 18: 513-516.

Chen, S.J. and Hwang, C.L. (1992). *Fuzzy Multiple Attribute Decision-Making: Methods and Applications*, Berlin: Springer-Verlag.

Cheng, R.C.H. and Amin, N.A.K. (1983). Estimating parameters in continuous univariate distributions with a shifted origin. *Journal of the Royal Statistical Society, Series B*, 45, pp. 394-403.

Chernozhukov, V. and Hong, H. (2002). Likelihood Inference in a Class of Non regular Econometric Models. *MIT Department Working Paper*, No. 02-05.

Clark, C.E. (1962). The PERT model for the distribution of an activity. *Operations Research*, 10, pp. 405-406.

Clemen, R.T. and Reilly, T. (2001). *Making hard decisions with decision tools*. Pacific Grove, CA: Duxbury.

Cooke, R.M. (1991). *Experts in Uncertainty*. New York, NY: Oxford University Press.

Couch , K. and Daly, M.C. (2000). Black-White inequality in the 1990's: a decade of Progress, *Working Papers in Applied Economic Theory*. No. 2000-07, Federal Reserve Bank of San Francisco.

David, F.N. (1962). *Games, Gods and Gambling*. London, UK: Charles Griffin.

Davidson, L.B. and Cooper, D.O. (1980). Implementing effective risk analysis at Getty oil company. *Interfaces*, 10: 62-75.

Davis, C. and Sorenson K. (Editors) (1969). *Handbook of Applied Hydraulics* (3-rd ed.). New York, NY: McGraw-Hill.

DeBrota, D.J., Roberts, S.D., Dittus, R.S. and Wilson, J.R. (1989). Visual interactive fitting of bounded Johnson distributions. *Simulation*, 199-205.

DeGroot, M. H. (1991). *Probability and Statistics*, 3rd ed. Reading, MA: Addison-Wesley.

Devore, J.L. (2004). *Probability and Statistics for Engineering and the Sciences* (6-th ed.). Toronto, Canada: Thomson, Brooks/Cole.

Donahue, J.D. (1964). *Products and Quotients of Random Variables and Their Applications*, Denver CO: Office of Aerospace Research.

Donald, S. G. and Paarsch, H. J. (1996). Identification, estimation, and testing in parametric empirical models of auctions within independent private values paradigm. *Econometric Theory*, 12: 517-567.

Draper, J. (1952). Properties of distributions resulting from certain simple transformations of the normal distribution. *Biometrika*, 39: 290-301.

Duffey, M.R. and Van Dorp, J.R. (1998). Risk analysis for large engineering projects: modeling cost uncertainty for ship production activities. *Journal of Engineering Valuation and Cost Analysis*, 2: 285-301.

Dutka, J. (1981). The Incomplete Beta Function - A Historical Profile. *Archive for History of Exact Science*, 24: 11-29.

Elderton, W.P. and Johnson, N.L. (1969). *Systems of Frequency Curves*. London: Cambridge University Press.

Engle, R. F. (1982). Autoregressive Conditional Heteroscedasticity with Estimates of the Variance of United Kingdom Inflations. *Econometrica*, 50: 987-1007.

Escher, M. C. (1989). *Escher on Escher*. New York, NY: Harry N. Abrams.

Everitt, B.S. and Hand, D.J. (1981). *Finite Mixture Distributions*. London, UK: Chapman and Hall.

Farebrother, R.W. (1990). Further details on contacts between Boscovich and Simpson in June 1760, *Biometrika*, 77: 397-400.

Fechner, G. T. (1897). *Kollectivemasslehre*. Leipzig: Wilhelm Englemann.

Flanigan, M.A. (1993). *A flexible, interactive, graphical approach to modelling stochastic input processes*. PhD Dissertation, School of Industrial Engineering, Purdue University, West Lafayette, Indiana.

Flehinger, B.J. and Kimmel, M. (1987). The natural history of lung cancer in periodically screened population. *Biometrics*, 43: 127-144.

Fullman, R.L. (1953). Measurement of particle size in opaque bodies. *Journal of Metals*, 5: 447-452.

Garvey, P.R. (2000). *Probability Methods for Cost Uncertainty Analysis, a Systems Engineering Perspective*. New York, NY: Marcel Dekker.

Griffiths, R.C. (1978). On a bivariate triangular distribution. *Austrian Journal of Statistics*, 20(2): 183-185*.

Grubbs, F.E. (1962). Attempts to validate certain PERT statistics or a 'Picking on PERT'. *Operations Research*, 1962, 10, pp. 912-915.

Gupta, A.K. and Nadarajah, S. (2004). *Handbook of Beta Distribution and its Applications*. New York, NY: Marcel Dekker.

Hald, A. (1990). *A History of Probability and Statistics and Their Applications before 1750*. New York, NY: Wiley.

Hall, P. (1932). The distribution of means for samples of size N drawn from a population in which the variates takes values between 0 and 1. *Biometrika*, 19: 240-249.

Harris, B. (1966). *Theory of Probability*. MA, Reading: Addison Wesley.

Hill, I. D., Hill, R. and Holder R.L. (1976). Fitting Johnson curves by moments. *Applied Statistics*, 25: 180-189.

Irwin, J.D. (1932). On the frequency distribution of the means of samples from a population having any law of frequency. *Biometrika*, 19: 234-239.

Johnson, D. (1997). The triangular distribution as a proxy for the beta distribution in risk analysis. *The Statistician*, 46 (3), pp. 387-398.

Johnson , N.L. (1949). Systems of frequency curves generated by the method of translation. *Biometrika*, 36:149-176.

Johnson, N. L. and Kitchen, J.O. (1971a). Some notes on tables to facilitate fitting S_B curves. *Biometrika*, 58: 223-226.

Johnson, N. L. and Kitchen, J.O. (1971b), Tables to facilitate fitting S_B curves II: both terminals known. *Biometrika*, 58: 657-668.

Johnson, N.L., Kotz S. and Balakrishnan N. (1994). *Continuous Univariate Distributions - 1*, (2nd ed.). New York, NY: Wiley.

Johnson, N.L. (1949). Systems of frequency curves generated by the methods of translation. *Biometrika*, 36: 149-176.

Johnson, N.L., Kotz S. and Balakrishnan N. (1995). *Continuous Univariate Distributions - 2*, (2nd ed.). New York, NY: Wiley.

Johnson, N.L. and Kotz, S. (1999). Non-smooth sailing or triangular distributions revisited after some 50 Years. *The Statistician*. 48 (2), pp. 179-187.

Kanefsky, M. and Thomas, J.B. (1965). On polarity detection schemes with non-Gaussian inputs. *Journal of the Franklin Institute*, 280 (2): 120-138.

Kapteyn, J.C. (1903). *Skew Frequency Curves in Biology and Statistics*. Astronomical Laboratory, Groningen, The Netherlands: Noordhoff.

Klein, G.E. (1993). The sensitivity of cash-flow analysis to the choice of statistical model for interest rate changes (with discussions). *Transactions of the Society of Actuaries*, XLV, pp. 79-186.

Kamburowski, J. (1997). New validations of PERT times. *Omega, International Journal of Management Science*, 25 (3) pp. 323-328.

Keefer, D.L. and Bodily, S.E. (1983). Three-point approximations for continuous random variables. *Management Science*, 29 (5): 595- 609.

Keefer, D.L. and Verdini, A.V. (1993). Better estimation of PERT activity time parameters. *Management Science*, 39 (9): 1086 - 1091.

Kelton, D.W., Sadowski, R.P. and Sadowski, D.A. (2002). *Simulation with Arena*. New York, NY: McGraw-Hill.

Kimmel, M. and Gorlova, O.Y. (2003). Stochastic models of progression of cancer and their use in controlling cancer-related mortality. *International Journal of Applied Mathematics and Computer Science*, 13 (3): 279-287.

Kleiber, C. and Kotz, S. (2003). *Statistical Size Distributions in Economics and Actuarial Sciences*. New York, NY: Wiley.

Kottegoda N.T. (1987). Fitting Johnson S_B curve by the method of maximum likelihood to annual maximum daily rainfalls. *Water Resources Research*: 23 (4):728-732.

Kotz, S. and Johnson N.L. eds. (1985). Moment ratio diagrams. *Encyclopedia of Statistical Sciences*, Vol. 5, New York, NY: Wiley, pp. 602-604.

Kotz, S, Kozubowski, T.J. and Podgórski, K (2001). *The Laplace Distribution and Generalizations*. Boston, MA: Birkhäuser.

Kotz, S., Kozubowski, T.J., and Podgórski, K. (2002). Maximum likelihood estimation of asymmetric Laplace parameters, *Annals of the Institute of Statistical Mathematics*, 54(4), pp. 816-826.

Kotz, S. and Van Dorp, J.R. (2004). Uneven two-sided power distribution: with applications in econometric models. *To appear in Statistical Methods and Applications*.

Kozubowski, T.J. and Podgórski, K. (1999). A class of asymmetric distributions. *Actuarial Research Clearing House*, 1, pp. 113-134.

Krantz, S.G. (2002). *The Implicit Function Theorem, History, Theory, and Applications*. Boston, MA: Birkhauser.

Kudus, A.K., Ahmad, M.I. and Lapongan, J. (1999). Nonlinear regression approach to estimating Johnson S_B parameters for diameter data. *Canadian Journal for Research*, 29: 310-314.

Lau, H., Lau, A.H. and Ho, C. (1998). Improved moment-estimation formulas using more than three subjective fractiles. *Management Science*, 44 (3): 346-350.

Law, A.M. and Kelton, W.D. (1991). *Simulation Modeling and Analysis*. New York, NY: McGraw Hill.

Leemis, L.M. (1986). Relationships among common univariate distributions. *The American Statistician*, Vol. 40, No. 2, pp. 143–146.

Leunberger, D.G. (1998). *Investment Science*. New York, NY: Oxford University Press.

Ljung, G.M. and Box, G.E.P. (1978). On a measure of lack of fit in time series models. *Biometrika*, 65, pp. 67-72.

Lloyd, E.H. (1962). Right Triangular Distribution. Chapter 29 in *Contributions to Order Statistics* (A.E. Sarhan and B.G. Greenberg (eds). New York, NY: Wiley*.

Lowan, A.N. and Ladermanm J. (1939). On the distribution of errors in N-th tabular differences. *Annals of Mathematical Statistics*, 10:360-364.

MacCrimmon, K.R. and Ryavec, C.A. (1964) An analytic study of the PERT assumptions. *Operations Research*, 12: 16-38.

Mage, D. T. (1980). An Explicit Solution for SB Parameters Using Four Percentile Points, *Technometrics*, 22: 247-251.

Malcolm, D.G., Roseboom, C.E., Clark, C.E. and Fazar, W. (1959). Application of a technique for research and development program evaluation. *Operations Research*, 7, pp. 646-649.

Mielke, P.W. Jr. (1975). Convenient beta distribution likelihood techniques for describing and comparing meteorological data, *Journal of Applied Meteorology*, 14 (6) pp. 985-990.

Mitra, S.K. (1971). On the probability distribution of the sum of uniformly distributed random variables. *SIAM Journal of Applied Mathematics*, 20 (2): 195-198.

Moder, J.J. and Rodgers, E.G.(1968). Judgment estimate of the moments of PERT type distributions. *Management Science*, 15 (2): B76-B83.

Mood, A., Graybill A.F., and Boes, D.C. (1974). *Introduction to the Theory of Statistics* (3rd ed.), Singapore: McGraw-Hill.

Nadarajah, S. (1999). A polynomial model for bivariate extreme value distributions, *Statistics and Probability Letters*, 42, pp. 15-25.

Nadarajah, S. (2002). A comment on van Dorp, J.R. and Kotz, S. (2002). The standard two sided power distribution and its properties: with applications in financial engineering (The American Statistician 2002 56(2):90-99). *The American Statistician* 56(4):340-41*.

Nadarajah, S. and Kotz, S. (2003). Moments of some J-shaped distributions, *Journal of Applied Statistics*, 30 (3): 311-317.

Nakao A. and Iwaki M. (2000). RBS study on Na-implanted polystyrene at various doses. *Applied Physics A, Materials and Science Procession*, 71 (2): 181-183.

Neuber, J.C. (2000). Axial burnup profile modeling and evaluation. *Second PIRT Meeting on BUC*, Washington, D.C. : jcn2000-08-14.

O'Neill, D., Sweetman, O. and Van de Gaer, D. (2002). Estimating Counterfactual Densities: An Application to Black-White Wage Differentials in the US. *Economics Department Working Paper Series*, Department of Economics, National University of Ireland - Maynooth.

Ostle, B., Prairie, R.R, Wiesen, J.M. (1961). Distribution of the range and midrange when sampling from a negatively skewed, right triangle population. *Industrial Quality Control*, 17 (9): 13-15.

Pairman E. and Pearson, K. (1919). On corrections for the moment-coefficients of limited range frequency distributions when there are finite or infinite ordinates and any slopes at the terminals of the range. *Biometrika*, 12 (3/4): 231-258.

Parks C.V., DeHart M.D. and Wagner J.C. (2000). *Review and Prioritization of Technical Issues Related to Burnup Credit for LWR Fuel.* Washington, D.C.: U.S. Nuclear Regulatory Commission, NUREG/CR-6665.

Patil, G.P., Boswell, M.T., Ratnaparkhi, M.V. (1984). Volume 2: Continuous Univariate Models, *Dictionary and Classified Bibliography of Statistical Distributions in Scientific Work, Vol. 7.* Burtonsville, MD: International Co-operative Publishing Hourse.

Pearson E.S. (ed) (1978). *The History of Statistics in the 17th and 18th Centuries against the Changing Background of Intellectual, Scientific and Religious Thought: Lectures by Karl Pearson given at University College, 1921-1933.* London: Griffin.

Pearson, E.S. and Tukey, J.W. (1965). Approximate means and standard deviations based on distances between percentages points of frequency curves. *Biometrika,* 52 (1965): 533-546.

Press, W.H., Flannery, B.P., Teukolsky, S.A. and Vettering, W.T. (1989). *Numerical Recipes in Pascal.* Cambridge, UK: Cambridge University Press.

Pouliquen, L.Y. (1970). Risk analysis in project appraisal. *World Bank Staff Occasional Papers, 1.* Baltimore, Md.: John Hopkins University Press.

Powell M.R. and Wilson J.D. (1997). Risk assessment for national natural resource conservation programs, *Discussion Paper 97-49.* Washington D.C.: Resources for the Future.

Pulkkinen, U. and Simola, K. (2000). *An Expert Panel Approach to Support Risk-Informed Decision Making.* Sateiluturvakeskus (Radiation and Nuclear Safety Authority of Finalnd STUK), Technical report STUK-YTO-TR 129, Helsinki Finland.

Rider, P.R. (1963). Sampling from a Triangular Distribution.*Journal of the Mathematical Statistical Association,* 58 (302): 509-512.

Runnenburg, J. Th. (1978). Mean, median, mode. *Statistica Neerlandica,* 32: 73-80.

Samuel, P. and Thomas, P.Y. (2003). Estimation of the parameters of triangular distributions by order statistics. *Calcutta Statistical Association Bulletin,* 54: 213-214*.

Schmeiser, B.W. and Lal, R. (1985). A five-parameter family of distributions. *Technical Report 84-7.* West Lafayette IN: School of Industrial Engineering, Purdue University.

Schmidt, R. (1934). Statistical analysis of one-dimensional distributions. *Annals of Mathematical Statistics,* 5: 30-43.

Selvidge, J.E. (1980). Assessing the extremes of probability distributions by the fractile method. *Decision Sciences,* 11 (1980): 493-502.

Seal, H.L. (1949). Historical development of the use of generating functions in probability theory. *Mitt. Vereinigung Scweiz. Versicherungsmathematiker,* 49: 209-228.

Sentenac D., Shalaginov A.N., Fera A. and de Jue W.H. (2000). On the instrumental resolution in X-ray reflectivity experiments. *Journal of Applied Crystallography,* 33: 130-136.

Siekierski, K. (1992). Comparison and evaluation of three methods of estimation of the Johnson S_B distribution. *Biometrical Journal,* 34 (7): 879-895.

Simpson, T. (1755). A letter to the Right Honourable George Earls of Maclesfield. President of the Royal Society, on the advantage of taking the mean of a number of observations in practical astronomy. *Philosopical Transactions*, 49 (1): 82-93.

Simpson, T. (1757). An attempt to show the advantage arising by taking the mean of a number of observatios in practical astronomy. *Miscellaneous Tracts on some curious and very interesting Subjects in Mechanics, Physical Astronomy and Speculative Mathematics*. 64-75.

Slifker, J. F. and S. S. Shapiro (1980). The Johnson System: Selection and Parameter Estimation. *Technometrics*, 22: 239-246.

Soofi, E. S., and Retzer, J.J. (2000). Information indices: unification and applications, *Journal of Econometrics*, 107, pp. 17-40.

Stigler, S.M. (1984). Studies in the history of probability and statistics. XL. Boscovich, Simpson and a 1760 manuscript note on fitting a linear relation. *Biometrika* 71 (3): 615-620.

Stigler, S.M. (1986). *The History of Statistics*. Cambridge, MA: The Belknap Press.

Straaijer A. and De Jager R. (2000). *EP 0 829 036 B9*. Paris: European Patent Office.

Stuart, A. and Ord, J.K. (1994). *Kendall's Advanced Theory of Statistics*, Vol 1., Distribution Theory, New York, NY: Wiley.

Stuart A., Ord J. K. and Arnold S. (1999). *Kendall's Advanced Theory of Statistics, Volume 2A: Classical Inference and the Linear Model (6th ed.)*. London: Edward Arnold.

Swain, J., Venkatraman, S. and Wilson, J. (1988). Least squares estimation of distribution functions in Johnson's translation system. *Journal of Statistical Computing and Simulation*, 29:271-297.

Taggart, R. (1980). *Ship Design and Construction*. The Society of Naval Architects and Marine Engineers (SNAME), New York.

Todhunter, I. (1865). *A History of Mathematical Theory of Probability from the time of Pascal to that of Laplace*. London: Macmillan.

Topp, C.W., and Leone, F.C. (1955). A family of J-shaped frequency functions, *Journal of the American Statistical Association*, 50 (269): 209-219.

Tsay, R.S. (2002). *Analysis of Financial Time Series*, New York, NY: Wiley, p.25.

Văduva, I. (1971). Computer generation of random variables and vectors related to PERT problems. In: *Proc. 4th. Conf. Probability Theory, Brasov* (ed. B. Bereanu), pp. 381-395. Bucharest: Editura Academiei Republicii Socialiste Romänia.

Van Dorp, J.R. and Mazzuchi, T.A. (2000). Solving for the parameters of a beta distribution under two quantile constraints, *Journal of Statistical Computation and Simulation*, 67, pp. 189-201.

Van Dorp, J.R., and Kotz, S. (2002a). The standard two sided power distribution and its properties: with applications in financial engineering, *The American Statistician*, 56 (2), pp. 90-99.

Van Dorp, J.R., and Kotz, S. (2002b), A novel extension of the triangular distribution and its parameter estimation, *The Statistician*, 51 (1), pp. 63-79.

Van Dorp, J.R., and Kotz, S. (2003a). Generalized trapezoidal distributions. *Metrika*, 58 (1): 85-97.

Van Dorp, J.R., and Kotz, S. (2003b). Generalizations of two sided power distributions and their convolution. *Communications in Statistics: Theory and Methods*, 32 (9): 1703 – 1723.

Van Dorp, J.R. (2004). Statistical dependence through common risk factors: with applications in uncertainty analysis, to appear in *European Journal of Operations Research*.

Van Dorp, J.R., and Kotz, S. (2004a). An attempt to resolve a PERT controversy. *Unpublished Manuscript*, Washington D.C: The George Washington University.

Van Dorp, J.R., and Kotz, S. (2004b). Modeling Income Distributions using Elevated Distributions on a Bounded Domain. *Unpublished Manuscript*, Washington D.C: The George Washington University.

von Seggern, D.H. (1993). *CRC Standard Curves and Surfaces*. Boca Raton, FL: CRC Press, pp. 124.

Vose D. (1996). *Quantitative Risk Analysis, A Guide to Monte Carlo Simulation Modeling*. New York: NY: Wiley.

Wagner, J.C. and DeHart M.D. (2000). *Review of Axial Burnup Distributions for Burnup Credit Calculations*. Oak Ridge, TN: Oak Ridge National Laboratory, ORNL/TM-1999/246.

Wagner, M.A.F. and Wilson, J.R. (1996). Using univariate Bézier distributions to model simulation input processes. *IIE transactions*, 28: 699-711.

Weibull W. (1939). A statistical distribution of wide applicability. *Journal of Applied Mechanics*, 18: 293-297.

Weyl, H. (1952). *Symmetry*. Princeton, NJ: Princeton University Press.

Wheeler, R.E. (1980). Quantile estimators of Johnson curve parameters. *Biometrika*, 67: 725-728.

Williams, T.M. (1992). Practical use of distributions in network analysis, *Journal of Operations Research Society*, 43, pp. 265-270.

Winston, W.L. (1993). *Operations Research, Applications and Algorithms*. Pacific Grove, CA: Duxbury Press.

Zabell, S. L. (1988). Symmetry and its discontents. In: B. Skyrms & W.L. Harper (eds) *Causation, Chance and Credence*, 1. Dordrecht: Kluwer, pp. 155-190.

Zadeh, L.A. (1965). Fuzzy sets. *Information and Control*, 8: 338-353.

Zhou, B. and McTague, J.P. (1996). Comparison and evaluation of five methods of estimation of the Johnson system parameters. *Canadian Journal for Research* 26: 928-935.

Author Index

The author index only includes the last name of those authors that appear in the text. In the case that authors have the same last name their initials have been added.

Subject Index